D0163107

AIRLINE OPERATIONS AND DELAY MANAGEMENT

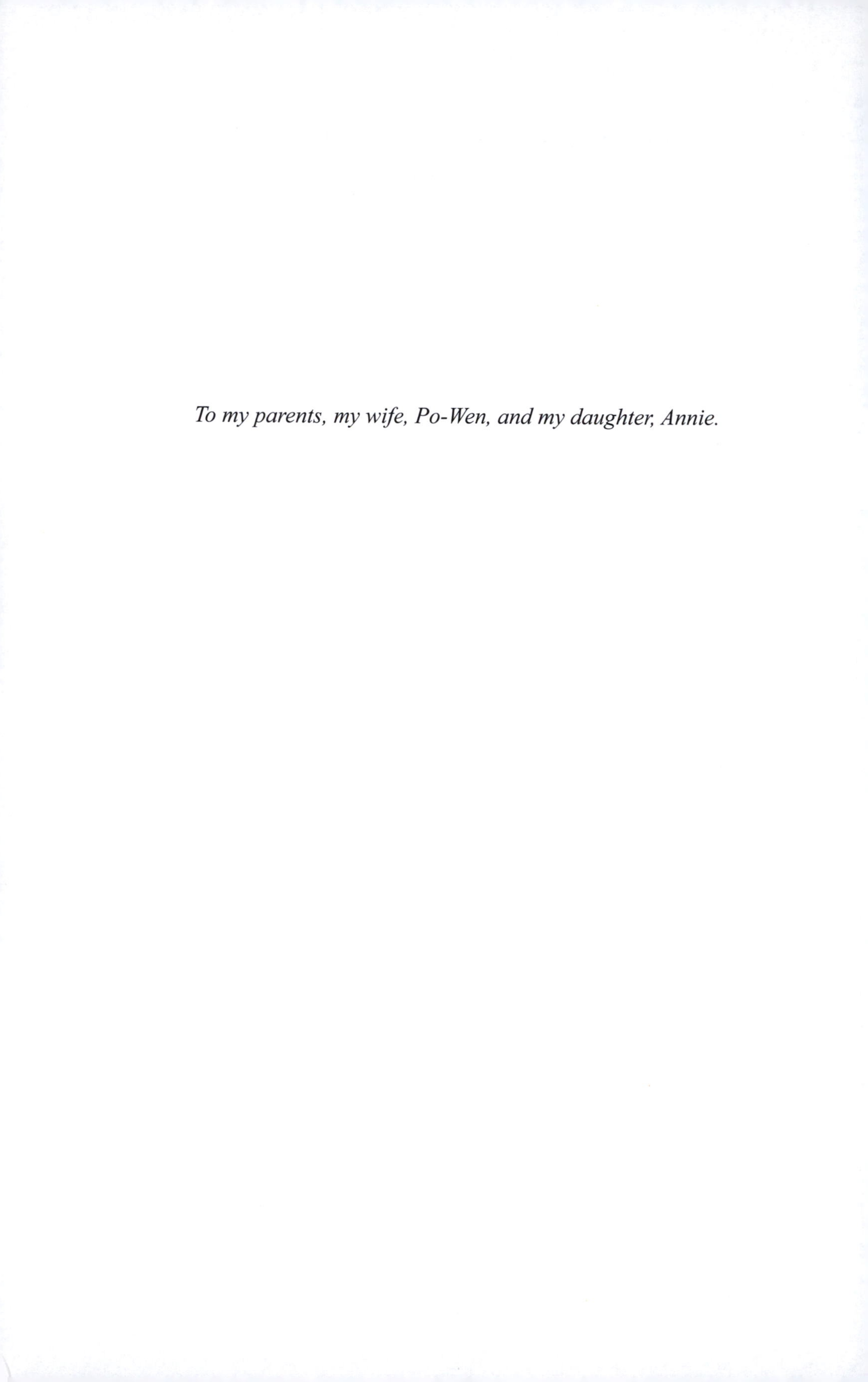

To my parents, my wife, Po-Wen, and my daughter, Annie.

Airline Operations and Delay Management

Insights from Airline Economics, Networks and Strategic Schedule Planning

CHENG-LUNG WU
Department of Aviation,
University of New South Wales, Sydney Australia

ASHGATE

Published by
Ashgate Publishing Limited
Wey Court East
Union Road
Farnham
Surrey, GU9 7PT
England

Ashgate Publishing Company
Suite 420
101 Cherry Street
Burlington
VT 05401-4405
USA

www.ashgate.com

British Library Cataloguing in Publication Data
Wu, Cheng-Lung.
Airline operations and delay management : insights from
airline economics, networks and strategic schedule planning.
1. Airlines--Management. 2. Airlines--Cost of operation.
I. Title
387.7'068-dc22

ISBN: 978-0-7546-7293-7 (hbk)
ISBN: 978-0-7546-9190-7 (ebk)

Library of Congress Cataloging-in-Publication Data
Wu, Cheng-Lung.
Airline operations and delay management : insights from airline economics, networks, and strategic schedule planning / by Cheng-Lung Wu.
p. cm.
Includes bibliographical references and index.
ISBN 978-0-7546-7293-7 (hardback) -- ISBN 978-0-7546-9190-7 (ebook) 1. Airlines--Management. 2. Business logistics. 3. Scheduling. 4. Operations research. 5. Airlines--Timetables. 6. Aeronautics, Commercial--Passenger traffic. 7. Strategic planning. I. Title.
HE9780.W82 2009
387.7068'5--dc22

2009032357

Reprinted 2011

Printed and bound in Great Britain by the
MPG Books Group, UK

Contents

List of Figures

List of Tables

Preface

If one has a chance to visit the Operations Control Centre of an airline and talk to a senior controller about the "ideal" airline schedule, they will surely describe some attributes of an ideal schedule that they would like to see in operations. If one is lucky enough to speak to a senior scheduler in an airline and ask the same question, they will definitely sketch a schedule which often looks different in many ways to that of an operations controller. After researching and consulting in airline operations for ten years, I have observed such internal conflicts in nearly every airline I have worked with in the past, regardless of the size of the company. Such conflicts are still seen in the industry now and they often result in operational losses to an airline in the forms of delays, schedule disruptions (and recovery), and un-satisfied passengers. The goal of this book is to bring these two diverging functions of an airline back together and work for the company's goal, i.e. profitability.

This book aims to provide readers with knowledge of both the practical and theoretical aspects of airline operations and delay management, as well as offering mathematical models for solving operational and schedule planning issues. The subject, *airline operations and delay management*, is explored extensively from a wide range of perspectives including: airline economics, airline business models, network development strategies, scheduling practices, and difficulties in actual airline operations. There is a strong need to address airline operations in the context of these wider views because many operational issues and problems are "inherent" in strategic airline schedule planning, and are inevitably encountered in actual operations, although operational issues are mostly manageable (with extra operating costs). This book adopts a different approach from conventional scheduling practices and explores the area of airline operations by addressing the inter-dependency between network development, economic/commercial driving forces in strategic airline scheduling, uncertain flight demands, and operational complexity of a modern airline network.

I have written this book in such a structure that it can serve as an introductory text for studying airline operations in a senior undergraduate course in a University. This book is also intended for those readers who would like to have a deeper understanding of contemporary schedule planning, airline operations, and delay management practices in the industry. Most of the contents of this book are written in a style such that concepts can easily be followed by professional analysts and aviation students even with limited aviation knowledge; some sections, in particular those developing mathematical models in various chapters, are intended more for those who pursue technical modelling knowledge and analytical tools for solving industrial problems such as aviation analysts, consultants, and postgraduate students in a University.

This book has been organised to allow individual chapters to independently address certain aspects of airline operations and scheduling, while as a whole the chapters provide a steady development when read from the beginning to the end. This book starts by introducing the operating environment of airlines, the evolution of airline networks, the driving economic forces behind airline scheduling, and the general practices of airline schedule planning procedures. Airline operations at airports are discussed in detail in Chapter 2, including activities by airlines at airports, the uncertainties involved in daily operations, and the impact of uncertain disruptions on the management of airline operations. Chapter 3 explores the issues in managing daily aircraft turnaround operations, combining industry practices and mathematical models for managing airline ground operations at airports.

The network effects of airline operations in Chapter 4 outline the impact of stochastic disruptions on a network scale, followed by the operational management of complex airline networks and on-time performance of a schedule in Chapter 5. Chapter 6 begins by introducing the emerging concept of robust airline scheduling and operational reliability of airline networks. Recent advances in modelling and optimising airline operations and schedules are discussed and examples are given in the book from real-world cases, wherever possible.

Preparing this book has occupied most Thursdays and Fridays of my work schedule since late 2007. Along the journey of writing, Prof. Jason Middleton at the Department of Aviation (UNSW) has provided me with the much-needed resources for book writing, in particular time and encouragement. I would also like to thank those graduate students in my research group, who have shown understanding to the fact that their research supervisor was working on his "second PhD thesis". Among my students, Dr. Theo Koo contributed to an earlier version of Section 3.5 of Chapter 3 on passenger flow management and passenger consumption behaviour at airport terminals. Prof. Jinn-Tsai Wong (NCTU Taiwan) contributed to the discussion on the modelling framework of delay propagation in Chapter 4 during his sabbatical visit in Sydney in 2006. Special thanks go to Miss Miriam Fewtrell who assisted in model development for Chapter 4 and editing the English of early drafts of this book. Just like cycling in the *Tour de France*, I would not have been able to finish this work and enjoy the journey of book writing without the excellent "team car" trailing behind me with the company of my family aboard: my lovely wife, Po-Wen and my sweet little daughter, Annie. Thank you and I love you.

Dr. Cheng-Lung (Richard) Wu
UNSW Aviation, Sydney Australia

Chapter 1
Introduction to Airline Operations and the Operating Environment

This chapter provides a brief introduction to airline network design issues, the economic and regulatory forces driving airline network evolution, airline schedule planning procedures, airline operation issues, the greater environment in which airlines operate, and the complexity of managing an airline network. Section 1.1 briefly discusses different types of airline networks, and the driving economic forces that have changed airline scheduling and networking in the past two decades. Section 1.2 examines common airline schedule planning processes and some shortcomings in airline operational management that result from this scheduling paradigm. Also introduced in this section is the concept of the "complex network" that makes airline scheduling, resources allocation and operational management challenging.

Section 1.3 covers airline operations both on the ground at airports and en-route in airspace. Various activities and procedures are required during airline ground operations and the success or efficiency of executing ground operational plans largely determines the schedule execution results, i.e. flight on-time performance (OTP) and profitability. Section 1.4 discusses the greater environment in which airlines operate, including airports and airspace. This section follows earlier sections and provides links between airline operations, schedule planning, complex network design, and delicate schedule synchronisation. The last section, Section 1.5 summarises this chapter and sets up the context for the remainder of this book. Section 1.5 explores airline operations, airline scheduling philosophy, and some management issues by addressing internal and external factors (with respect to an airline) that significantly influence tactical airline operations and strategic airline planning. The outline of the remaining chapters of this book is provided in this section as well.

1.1 Airline Networks and Airline Economics

1.1.1 Airline Network Types

Broadly speaking, the term "airline network" has two meanings. First, an airline network means the network that is formed by those airports that are serviced directly or indirectly by an airline. This definition places an emphasis on network coverage (the "spatial" coverage) and the seamless services that an airline and its partner airlines can provide customers. The second meaning of an airline network focuses on the more "temporal" attributes of an airline network. The temporal attributes of an airline network are characterised by the frequency of services between two airports and the extent an airport is connected with other airports in the network via various service types (direct flights or transfer services) and service frequency. The strong focus and competition, both on the temporal and spatial attributes, gives rise to various airline network models. Among these, some focus more on the temporal attributes such as a hub-and-spoke network, and others focus more on the spatial attributes or market orientation such as a point-to-point network.

A hub-and-spoke network consists of at least one "hub airport" that plays the role of "collecting and distributing" passenger traffic among flights at an airport. Passengers fly to the hub airport and then either transit to another flight to reach their destination, or end their journey at the hub airport. To efficiently "exchange" transit passengers among flights at a hub, inbound flights come in "waves" (or "banks"), followed by waves of outbound flights. The role of a hub airport is to facilitate intensive and often dense aircraft ground operations during the time when passengers need to travel from one gate to another for a connecting flight. Since limited direct connections are available between spoke airports in a hubbing system, most spoke-to-spoke connections must be made via a hub, resulting in a longer travel time than a direct service between two spokes. In a large geographic region such as North America, there may be two or more hubs in a hubbing network to facilitate passenger connections.

A point-to-point network, on the other hand, focuses more on individual markets. Passengers are often able to travel directly between two airports which are serviced by an airline that operates a point-to-point network. Due to the nature of such a network system, it is not always possible for passengers at a city to travel to any other cities that are serviced in a point-to-point network, because the market between any two cities in the network may not sustain a direct flight. The emerging low-cost carriers (LCCs), pioneered by the Southwest Airlines in the U.S., take advantage of this type of network system, and this "low-cost" sector in the industry has expanded quickly around the world in different continents.

Naturally, the geographic location of the base airport of an airline has a profound influence on the network model that an airline adopts. This gives rise to the "mixed network" model that adopts both attributes from the hubbing system as well as from the point-to-point system. In such a network, the intensity of hubbing is not as high as that in a hubbing system and the focus of hubbing is often *ad hoc* and aimed at providing flight connection opportunities between specific markets. Hence, inbound and outbound flights at such a mixed hub are usually highly synchronised and for most hubs, this operation is often "directional" in terms of traffic. For instance, Qantas and British Airways operate a hub at Singapore Airport where there is north bound traffic to Europe in the evening and south bound traffic to Australia in the early morning.

The growth of airline alliances has also changed the definition of an airline network. Through code-sharing agreements among partner airlines in an alliance, an airline is able to expand its network by including airports/flights that are not directly operated by itself. This "virtual" expansion of an airline network provides airlines with economic benefits due to the lower costs of operating some code-sharing markets, whilst providing passengers with better network coverage.

1.1.2 Economic Forces

The "structure" of an airline network is the result of airline route planning, which is aimed at maximising network revenues by changing the structure, i.e. by adding and/or dropping airports or by changing service frequency between airports in a network. The driving force of airline route planning is deeply rooted in corporate revenue maximisation, although there are "flag carriers" that may operate certain routes more for political considerations than economical ones. In a highly regulated market such as the U.S. market in early 1970s, airlines operated certain routes for various considerations, e.g. national and community benefits, and there was usually limited competition in most markets. After deregulation in the U.S. market in 1978, market forces came into play and dramatically reshaped the network models of many American carriers. The most obvious change has been the emergence of the hub-and-spoke system after deregulation.

Due to geographic constraints and the nature of directional (east-west) traffic in North America, economic forces have driven the development of hub-and-spoke systems for decades. An airline that operates such a network can readily vary the frequency of services to a market, so as to easily extend flight supplies or to fend off new entrant carriers into the same market. The concentration of flights at certain hubs also creates the desirable "economies of density", meaning that the marginal cost of varying flight density (frequency) is relatively low in such a network structure, so an airline can easily generate more profits by increasing frequency (Button 2002). Pursuing this potential economic benefit, airlines started developing strong hubbing systems in the U.S. after deregulation.

Another emerging force in the deregulated market is to focus on origin-destination (OD) markets. This is the underlying economic rationale of most low-cost carriers: an OD market is added to the network if and only if the market is profitable and sustainable in the short run. In such a point-to-point network, the focus is less on the economies of density or the "network synergy", but more on the profitability of individual OD markets in the network. Potential connecting markets and network coverage are not the central issues of airline route planning. Hence, few connections ("on-line" connections in particular) are facilitated or available between markets, although passengers may still independently organise self-hubbing or self-connection at certain airports that serve more destinations in a region.

1.1.3 Complex Networks and Network Effects

Compared with other forms of network systems, an airline network is a "time-varying network", in which the "physical" links between nodes (i.e. airports) in such a network are formed by flights on a timetable. Since there are only a certain number of flights scheduled at specific times between a city pair in a network, the "physical structure" of an airline network depends on the times when flights are scheduled, hence being a time-varying network. Another unique attribute of an airline network, which is similar to many large complex network systems, is that most of elements in the network are subject to stochastic influence. For airports, the operation of an airport is subject to weather conditions, which influence the practical runway capacity and some ground operations. For airlines, ground operations are subject to uncertainties coming from external sources such as aircraft ground service delays and shortage of ground staff, and internal sources such as connecting crew delays. For airspace, uncertainties come from air traffic control loading in individual sectors, and congested terminal areas around large airports during peak hours.

The combined time-varying and stochastic nature of an airline network creates a set of "dynamics" that affect the way an airline operates and manages its operations. Like many other large complex networks, the operations at an airport may influence other parts of the network through traffic flows, e.g. aircraft routing and passenger connections. Constraints at a particular airport may cause partial network degradation. For instance, a thunderstorm around a major hub airport can cause delays to hundreds of flights and even flight cancellations across the network due to the capacity reduction of runways. Although economic forces drive airlines towards revenue maximisation and asset utilisation, operational issues arising in running a complex network may offset some of these financial gains. It is in the context of this conflict between potential financial gains and uncertainties in real-world operations that we will examine airline operations and scheduling issues. This will provide the overarching context of this book in general.

1.2 Airline Schedule Planning

1.2.1 Schedule Planning Stages

There are four major stages in airline schedule planning. First of all, *network planning* (also called route planning or schedule generation) is the fundamental step in schedule planning. The goal of network planning is to explore potential markets and forecast market demands for certain services, e.g. leisure demand or business demand. Hence, this stage of work involves demand modelling, market forecasting and initial schedule establishment. Airlines often start route planning well ahead of operations and often base this work on the existing network by adding/cutting services and optimising/changing frequency to capitalise on existing and new markets. The objective of airline network planning is to set up a preliminary timetable to generate maximum profits with limited resources such as aircraft fleets, capital investments and human resources.

The second stage of schedule planning is to conduct *fleet assignment* based on the outcome of route planning from stage one, i.e. the preliminary timetable. The goal of fleet assignment is to explore the best possible way to execute the timetable by assigning available fleets with the minimum operating costs. Since operating different types of aircraft would incur different levels of costs and provide different levels of seat capacities, individual demand forecasting on a market basis is essential at this stage, to match potential or expected market demand with the appropriate supply of seats (i.e. the right type of aircraft). Hence, the objective of fleet assignment is to "partition" the proposed timetable into a number of "flight sets" which are operated by different fleets in a way which minimises cost.

The third stage of scheduling is *aircraft routing*. For each "set" of flights obtained from fleet assignment, an aircraft routing plan is developed. Flights flown by a specific fleet are partitioned and "connected" in a chronological order to form a "route", such that an aircraft of this fleet can operate those flights in a route sequentially. The other objective of aircraft routing is to route each aircraft to specific maintenance stations according to aircraft maintenance requirements and schedules. The timely routing of an aircraft back to the maintenance station is essential for safety considerations.

The fourth stage of scheduling is *crew rostering*. Flight crew are specifically fleet type endorsed, so a pilot can only be rostered to fly a certain type of aircraft. Cabin crew are generally not constrained by this endorsement, but would require specific training for those aircraft types they will operate. Crew cost is usually the second largest cost item (after fuel cost) on the balance sheet of an airline, so crew rostering is a critical stage where an airline can potentially generate substantial savings. Hence, the goal of crew rostering is to utilise crew resources to execute the planned timetable at a minimal cost. Since the crew are humans and not like aircraft that can be on duty for extended hours, there are safety regulations on the workload of pilots by civil aviation authorities in different countries. In addition, crew unions often have enterprise bargain agreements with airlines that further limit the workload and conditions in which crew rosters can be designed. In a sense, crew rostering is similar to aircraft routing, but with additional restrictions on fleet types, working hours, and the location of crew bases in the network.

1.2.2 Resource Connections and Delicate Schedule Synchronisation

In addition to being time-varying and subject to stochastic disruptions, the complexity of an airline network is deeply rooted in the high levels of synchronisation among the four "layers" of a network – the aircraft routing network, crewing network, passenger itinerary network, and cargo shipping network. The *aircraft routing network* is the "backbone" of an airline network system, with which other networks/sub-systems are synchronised accordingly. Aircraft are routed to conduct a specific flight at a specific time, and meet the frequency demands between destinations. An important task in aircraft routing is to route the aircraft back to a maintenance base in time for the required maintenance services.

Once a specific type of aircraft routing network has been decided upon, crew (cabin and cockpit crew) are assigned to operate a number of flights during a certain period of time (called *crew pairing* or a *tour of duty*). Since the endorsement of cockpit crew must match the aircraft type, a group of cockpit crew and cabin crew may follow an aircraft for a number of flights during a day before switching to another aircraft or ending the duty of the day. This creates the layer of the *crew rostering network* (also known as the *crew pairing network* in the industry), which is highly synchronised with the movement of aircraft in the aircraft routing network.

In addition to these two networks, passengers take flights that transport them from origin airports to destination airports. If a journey involves a transfer at an airport then the whole "itinerary" of a journey will span across a number of flights in the network. Since there are millions of passengers travelling between airports in a network, there is a high degree of itinerary overlapping among passengers, especially for hub-bound traffic; the individual itineraries of passengers on a flight vary and "overlap" with one another at least for this flight. This layer of the *passenger itinerary network* is driven by market demands in the network and follows the nature of airline networks in that it is both time-varying and subject to stochastic disruptions.

The fourth layer of an airline network is the *cargo shipping network*. This layer is not as critical as the other three, because most cargo/mail shipments can tolerate delays (for non-express shipments), or can be routed through a number of airports before reaching the destination (depending on the destination of shipments and the location of cargo hubs in a network). For "belly cargo" (i.e. those cargo that are carried by passenger flights), they form a similar type of network to the passenger itinerary network. Freighter flights form a separate sub-network which contains a separate aircraft routing sub-network (due to different fleets from passenger aircraft), and a crew rostering sub-network (due to different crewing requirements and no passenger charter).

We can see from the structure of an airline network that the high levels of schedule synchronisation between layers contributes significantly, both to the complexity of network planning and also to the complexity of daily operational management in such a network. If an aircraft is delayed, passengers, crew and cargo aboard are also delayed. Late passengers and crew may also delay other flights down the line at other airports that are supposed to receive connecting passengers and crew. For major disruptions such as the closure of an airport due to a thunderstorm, there may be numerous inbound and outbound flights at the affected airport that need to be cancelled or delayed. This causes a ripple effect throughout various layers of an airline network due to the resource connections and synchronisation among them.

In the grid of airports, this ripple effect also causes delays at some other airports because of airline operations in the airport network. A highly synchronised airline network runs well in normal operations and is able to cope with minor stochastic disruptions. However, the conventional approach we have introduced to airline schedule planning may be vulnerable to stochastic forces and major disruptions, which unfortunately occur during airline operations. We will explore this issue in depth and how the industry has been addressing it recently in Chapters 5 and 6 of this book.

1.3 Airline Operations

1.3.1 Airline Operations at Airports

The goal of airline operations at airports is to facilitate the execution of airline schedules, movements of passengers, baggage, and cargo. Airline operations at airports can be generally categorised by activities on the landside and on the airside of an airport by an airline. On the landside, airline operations involve passenger check-in, baggage check-in, connecting passenger/baggage processing, cargo and goods handling (in a cargo processing facility adjacent to an airport), catering service preparation (usually in a facility nearby an airport and often next to the airside), and passenger boarding at the gates.

The area beyond the boarding gates belongs to the airside of an airport. Activities on the airside by airlines mainly include the tasks of turning around aircraft at the gates. This turnaround operation includes passenger handling (disembarkation and embarkation), cabin cleaning, crewing (crew change), routine visual maintenance checks, re-fuelling, cargo handling (unloading and loading), and catering services (unloading and loading). Some airlines have their own ground handling agents that carry out both airside and landside services, especially at their base or hub airports. Other airlines outsource ground handling services to independent service agents or to different airlines for cost-cutting purposes, especially at non-hub airports (also known as *out-stations* in the industry).

Airline operational activities at an airport, like those activities in some service and manufacturing industries, have unique characteristics including operating time constraints and a standardised operating procedure. The timeline of airline operations is based on the scheduled departure time of flights in the timetable. Passenger check-in starts around one and a half hours before departure for a domestic flight, and up to three hours for an international flight, depending on aircraft size. Aircraft turnaround operations are conducted within the scheduled turnaround time between two flights, which is fleet type and service requirement dependent. For a domestic low-cost service by a B737 or A320, the turnaround time can be as short as 15 to 20 minutes by some low-cost carriers. For a full-service by a B747 on an international route, the turnaround time can be between one and a half hours to two hours.

Aircraft turnaround activities are often standardised to a strict timeline, and most airlines follow their own standard operating procedures (SOPs) or the ones provided by aircraft manufacturers. For most activities, there are planned operating sequences in the SOP. For instance, check-in counters are open from two and a half hours before the scheduled departure time (for a wide body aircraft operation) and are closed thirty minutes before the departure time. For turnaround activities, on-board passengers disembark first, then followed by crew (if needed), cabin cleaning, crew changes (if needed), and passenger boarding.

Most workflows can run simultaneously with limited interference between them such as the passenger handling flow and the cargo/baggage handling flow. However, disruptions can occur to any activities in the workflows and delays to an activity may disrupt other activities and eventually delay a flight, due to the sequential nature of most workflows. In addition, workflow interference may occur among flights that are being turned around at the same airport at the same time, due to connections of passengers, crew and goods/baggage, i.e. synchronisation between flights. This type of connection can occur within flights operated by an airline or between flights by different airlines at the same airport.

1.3.2 Enroute Flight Operations in a Network

Enroute operations consist of the procedures after aircraft push-back at the gate of the departure airport and before the arrival at the gate of the destination airport (also called *gate-to-gate operations* in Europe). Enroute airline operations mostly involve airborne flight operations and aircraft operations on the ground at airports. For airlines, "flight operations" often refers to the management of flights when airborne. These operations are carried out by cockpit crew and are facilitated by air traffic controllers in the various sectors through which an aircraft will fly. There is a certain level of flexibility such that an airline can choose the "optimum" flight path of an airborne aircraft between an origin and destination, depending on enroute weather conditions and estimated fuel consumption.

The flight path an aircraft takes is often planned a few days before departure and is also subject to updates on the day of departure by the airline. Considerations to flight path planning (a part of *flight dispatch*) include the estimated fuel consumption along the path, engine numbers and emergency needs (also known as "ETOPS"), weather forecast along the path (influencing fuel consumption and safety), and the estimated take-off weight (influencing the fuel carried aboard). Delays occurred during flight operations are mostly due to requested detour by air traffic controllers for various reasons such as: weather conditions on the path, and congested air traffic control (ATC) sectors. The congestion in the terminal manoeuvring area (TMA) around the destination airport may force an arriving aircraft to join a landing queue in an airborne holding pattern and await a landing slot.

Aircraft enroute operations are largely beyond an airlines' control, with the exception that airlines are able to alter their flight plans in advance subject to the current practice of the air traffic control and management. Some airlines may be able to request a shuffle of departure or landing slots at an airport under specific circumstances such as inclement weather, or if they are the dominant (hub) carriers at the airport. However, this scenario only happens during heavily disrupted periods, when airlines need to cancel, delay and prioritise some flights in order to minimise the impact due to the unexpected schedule disruptions.

Aircraft operations on the ground of an airport are procedural, and in most cases are facilitated by the air traffic control services provided by airport towers. This sequence of operations at the departure airport involves the captain requesting for push-back permission, ground staff pushing back the aircraft from a gate to the taxiway, the aircraft taxiing on the taxiway, waiting in the departure queue at the holding apron of the runway end (if necessary), and taking off (wheels off the runway). The operation at the arrival airport follows an opposite but similar procedure from landing (wheels on the runway), taxiing on the taxiway, to parking at the arrival gate. Congestions for arrival aircraft may occur on the taxiway or before parking at a gate, due to late gate clearance.

1.4 The Operating Environment of Airlines

1.4.1 The Nature of Airline Scheduling and the Greater Environment

Airline operation is the execution of a complex airline schedule system which is designed to maximise corporate revenues in a specific market, which is formed by a network of airports. Schedule planning itself contains major strategic and tactical decisions made by an airline, with the aim of utilising available corporate resources and maximising revenues. Like other industries, the airline industry is a competitive market in which market forces such as supply, demand and pricing play a key role in airline strategic planning, scheduling and consequently schedule execution. The nature of an airline business is like any other businesses, but the operational side is more constrained by the "greater environment".

From an airline's perspective, an airline operates in a "man-made" environment, in which airports provide ground infrastructures, and air traffic control authorities provide air traffic management services (allocating limited capacity). Unlike the free market that most businesses or industries operate, this man-made environment imposes numerous constraints on airlines not only in terms of strategic planning, but also on daily operations. Strategically, the limited capacity of airports, bilateral air service accords between countries, and consequently limited air traffic services constrain the free growth of an airline in certain markets. Under such constraints, airlines cannot freely expand services to any markets that may have business potential.

Tactically speaking, airline operations are stochastic in nature when compared with the "fixed" schedule that an airline needs to operate. Some operations depend on weather conditions, while others may depend on workload that is often directly related to the actual number of passengers carried on the day of operation. Therefore, we need to have a good understanding of the "environment" in which airlines operate, before further elaborating on airline schedule planning and operational improvements. Interestingly, it is also the operating environment that influences or even enforces how airlines plan schedules and more importantly, how airline operations perform in such a highly constrained environment.

1.4.2 Complex Networks and Schedule Synchronisation

An airline network involves layers of sub-networks which inter-connect with each other via resources (aircraft and crew) or passengers. An airline network also connects with the networks of other airlines, mostly via connecting passengers in code-share operations or reciprocal bilateral agreements. Given the unique attributes of the complex aviation system, as discussed earlier, airlines operate in a unique environment in which delicately synchronised connections among networks take place at many airports in the network. Airline operation is expected to adhere to the planned schedule, though stochastic disrupting events can occur to any elements of the complex system and in some cases, may cause serious network-wide impacts.

The "network effect" on an airline system is a double-edged sword: on the one hand, airlines take advantage of networks to utilise resources and assets in order to maximise corporate profits; on the other hand, the complex web of resource synchronisation inherently contains some degrees of operational uncertainty that often lead to extra operating costs. The driving force of utilising airline resources and assets has resulted in schedule plans that are over-optimistic on the drawing board, or are optimised only for normal operations. With such a plan, the well-synchronised network runs smoothly under normal operating conditions. However, when the plan is disrupted, more resources are needed to restore the system back to its normal status. This scheduling philosophy has dominated airline scheduling for more than a decade. It is not until recently that airlines started to look for solutions which create "robust" airline schedules that can improve operational reliability in the actual environment that airlines face. We will explore how this shift of scheduling paradigm has evolved recently in Chapter 6.

1.4.3 Stochastic Operations and Disruptions

Stochastic forces affect airline scheduling and operations in two ways. First, many tasks in airline operations are naturally stochastic, in terms of the time required to finish those tasks. For instance, the time required to load baggage depends on the number of passengers on the day of operation, as well as on the actual weight of the baggage brought by all passengers aboard. Since the actual number of passengers aboard a flight is unknown until check-in is finished, the service time of a few tasks is always unknown until the last minute. Moreover, catering service time and fuel carriage also depend on the actual passenger number on the day of operation. This uncertainty brings a stochastic factor into airline operations.

Second, disrupting events in airline operations and in the operating environment (airports and airspace) occur in a stochastic nature, such that many of disruptions are not predictable. Major events such as aircraft technical issues and severe weather conditions are hard to prepare for and the consequences of major disruptions like these are always serious and expensive for airlines and the society. Minor disruptions such as delays and late inbound passengers/bags have less of an impact, but gradually, delays may accumulate and propagate across some parts of a network, due to the synchronisation in an airline network. The effect of delays in a network will be further modelled and discussed in detail in Chapters 4 and 5.

1.5 The Context and Outline of the Book

In this book, we will focus on airline operations and some management issues specifically pertaining to operations such as delay management, operating process optimisation, schedule optimisation, and schedule disruption management. Since airline operation is subject to stochastic forces and is a result of airline schedule planning, we will approach airline operations from the perspective of airline schedule planning and optimisation. Thus, we do not only address the "consequences" of airline operations in this book, we also explore the "root causes" resulting from airline scheduling that are often overlooked by airline management. Further up the scale, we shall bear in mind that the philosophy of airline schedule planning is deeply rooted in economic principles and market forces, some of which are imposed or constrained by the operating environment of this industry. Therefore, many operational issues, indeed are "created" by airlines themselves, because of the constraints from the environment (airports and airspace), the risk of pursuing potential economical benefits, and uncertainties naturally inherent in a complex system. It is in this overarching environment and unique context that we will examine airline schedule planning, airline operations, operational control, and the evolving schedule planning philosophy in this book.

The outline of the remaining book is to focus on airline operations on the ground in Chapters 2 and 3. Chapter 2 explores the influence of airport constraints on airline scheduling and operations on the ground. Apart from limited airport slots for departures and arrivals, airlines are subject to the time constraint imposed by their own planned schedules to conduct aircraft turnaround operations. An analytical model that considers uncertainties in airline operations is provided and we shall run a few scenario tests to explore the extent to which airline operations are subject to uncertainties and self-imposed scheduling considerations. Chapter 3 shifts the focus to some operational issues, in particular: delay management, delay data collection, managing aircraft turnaround operations and managing passenger flows in airport terminals. A micro-simulation model is developed as a tool that allows us to examine in detail how individual service processes of turning around an aircraft are synchronised with each other, and how collectively they can affect flight punctuality.

Having explored airline operations, the book moves on to discuss enroute flight operations by airlines in Chapter 4 and airline disruption management in Chapter 5. Chapter 4 focuses on the "network factor" in airline scheduling and operations, and extends the scope of this book from a "local perspective" (i.e. at an individual airport level) to a network and system perspective. This chapter starts by introducing aircraft routing, fleeting, crewing requirements, then crew scheduling. Next, an airline network is broken down into sub-networks with which we explain the delicacy of schedule synchronisation and resource connections in this complex system. Based on this network view, we return to the topic of delay management and explore the potential impacts of delays on a network scale, namely delay propagation.

Chapter 5 expands on this by addressing some issues in daily airline operational management such as disruption management, and on-time performance reporting. A case study based on a set of real data is provided to demonstrate a common approach by airlines to delay and operation diagnosis. Building upon the network view in Chapter 4, we introduce operational uncertainties in the network and the concept of "inherent delays" that results from airline scheduling philosophy/policies, network features, and constraints imposed by the operating environment, e.g. capacity limitations of airports. A schedule optimisation model is then presented to demonstrate the possibility of improving the robustness in airline operations by an emerging scheduling concept, namely robust airline scheduling.

Chapter 6 will firstly provide a brief review of past practices of scheduling in the industry. Subsequent to this review, this chapter introduces the concept of schedule robustness and how this concept is currently evolving in the airline industry. Some strategic and tactical methodologies are discussed in reference to improving schedule robustness in strategic schedule planning and tactical operations. Before finishing this chapter and concluding this book, we return back to the debate of airline schedule planning and its on-going struggle with the complexities of network synchronisation and market-driven economic forces in corporate finance. We lay out some potential scheduling philosophies for the future based on the emerging robust scheduling and integrated modelling concepts.

Chapter 2
Airline Operations at Airports[1]

Chapter 2 details the activities involved in airline operations at airports. Although for most activities of airline operations at airports there are "standard procedures", most airlines adopt these procedures with some in-house modifications, aiming at improving operational efficiency. Accordingly, details about these "in-house modified operational procedures" are often not available in the public domain and are treated as commercial in confidence by airlines. Hence, this chapter covers those standard procedures of airline operations at airports, while introducing some ad hoc operations of airlines wherever data are available from public sources. Operations at airports may also differ between network carriers and low-cost carriers. Differences will be identified and detailed where necessary and relevant. Mathematical models developed to improve the operational efficiency of airline ground operations are described in detail.

This chapter begins by introducing airline scheduling philosophy and its implications on airline operations for different types of airline networks, e.g. hubbing and point-to-point services. The impact of airline scheduling on the allocation and availability of airport slots and airport infrastructures is discussed in detail in Section 2.1. Section 2.2 further delineates operational activities involved in aircraft turnaround operations, including passenger processing and goods handling. Section 2.3 moves on to discuss the stochastic variations of service time of ground operations, limited turnaround time in schedules, and random flight delays.

Based on these discussions, Section 2.4 introduces an empirical method which is commonly used by airlines to deal with uncertain delays and pursue the optimal allocation of turnaround time. Section 2.5 continues this by developing an analytical method that provides sufficient modelling details to meet scheduling demand. The last section is a case study based on real airline data which demonstrates some key concepts regarding punctuality management and airline scheduling. Within these discussions, modelling concepts are introduced gradually as relevant to the content. The pursuit of efficient and effective allocation of turnaround time in airline scheduling is discussed with the presentation of mathematical models and applications in both real and hypothetical cases.

---

1 This chapter is partially based on the following publications: Wu and Caves (2000); (2002); (2003); (2004).

2.1 Airline Scheduling at Airports

The "philosophy" of airline scheduling differs between airlines and the development of the "philosophy" is strongly influenced by two key forces in scheduling, namely commercial interests and technical considerations. In the next two sections, we will explore how these two forces shape the philosophy of airline scheduling and how different scheduling philosophies influence airline business models and operations.

2.1.1 Hub Scheduling and Operations

Commercial interests are driven by the desire to maximise revenues of airline products, i.e. the flights on airline timetables. Flights are the "intangible and perishable" products that airlines offer customers. The "intangibility" reflects the fact that the consumed product is those "services" provided by airlines aboard aircraft to transport passengers from an origin airport to a destination airport. The "perishability" reflects that once a flight departs according to a timetable, the "product" no longer exists in the market and has perished immediately. Hence, the "attributes" of airlines products, e.g. onboard/ground services and flight timetables, determine the saleability of these products as well as the revenue of sales.

Within these attributes, airline scheduling philosophy is strongly influenced by the "perishability" of products and the desire to maximise the saleability of products by scheduling flights at the most attractive times for passengers. Generally speaking, there are three broad airline scheduling models and each reflects different scheduling philosophies and airline business models including: the hub-and-spoke model, the mixed hubbing model, and the point-to-point model.

A hubbing schedule is driven by the desire to "exchange" passengers between an inbound and an outbound "wave" of flights at a hub airport after a short period of connection time. This type of hubbing networks creates numerous connection options among airports of a network and also creates the desired "economies of density" for hubbing airlines. A common hubbing schedule looks like the one in Figure 2.1 that has clear arrival peaks and departure peaks. This scheduling philosophy is ideal for a network in which there is regional and/or directional traffic, e.g. Munich for Lufthansa, Atlanta for United Airlines, Dallas-Fort Worth and Chicago O'Hare for American Airlines. A hub-and-spoke network is also ideal for domestic operations with a few "gateway airports" in the network serving as the hubs between domestic and international connections such as Los Angeles, New York JFK and Frankfurt. For other cases, the hub location may enjoy some geographical benefits for traffic exchanges, e.g. Singapore and London Heathrow Airport.

Hubbing schedules are not universally applicable for every airline. Those airlines which do not have enough destinations or demand in their networks to sustain frequent and dense wave structures will not enjoy the benefits of strong hubbing. However, the model of "weak hubbing" and "rolling hubs" provides an answer to this situation. Some carriers naturally adopt the "weak hubbing" model, mainly for directional and long-haul traffic with regional feeding traffic in a network, e.g. Qantas at Singapore, British Airways at London Heathrow Airport and Emirates at Dubai. Often, "weak hubbing" schedules comprise both hubbing functions and point-to-point services between spokes and hubs. In such a network, the hubbing function facilitates passenger interlining between flights and the point-to-point service provides a channel to feed regional/domestic traffic to hubs. The figure below (Figure 2.2) illustrates a possible traffic pattern by a "weak hubbing" carrier which operates hubbing flights in the early morning and in the evening. It is noted that flight frequency in a weak hubbing network is significantly lower and the number of waves is also less than a strong hubbing model, as seen earlier.

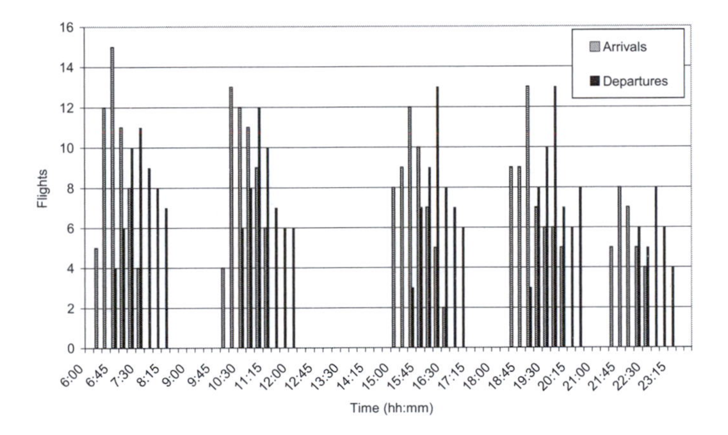

Figure 2.1 A strong hubbing schedule (hypothetical data)

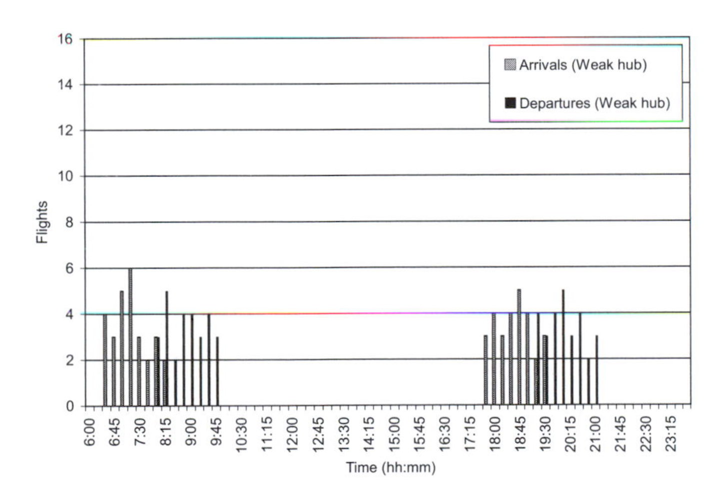

Figure 2.2 A weak hubbing schedule (hypothetical data)

The rolling hub model is a modified hubbing concept that came from the strong hubbing model. Although a strong hubbing network can create certain benefits, it also generates some hidden operating costs such as high delay costs in a network and low airport infrastructure utilistation (i.e. high asset costs). Since 2001, airlines have started reducing the intensity of flight hubbing by spreading the peak traffic of inbound and outbound waves more evenly during peak and off-peak hours of operations. As the peaks of inbound and outbound traffic are smoother, flight hubbing activities now do not only appear during the peak traffic period but more evenly spread out along the day, hence the name: *rolling hubs*. Rolling hubs still exhibit the structure of hubbing, but the hubbing "strength" is not as intense as the strong hubbing model. American Airlines and Lufthansa adjusted their networks and modified some hubbing airports to rolling hubs, e.g. O'Hare, Dallas-Fort Worth, and Frankfurt as shown in Figure 2.3, an example rolling hub structure. A noticeable difference between strong hubbing and a rolling hub is that a rolling hub has lower demand for peak slots and the duration of an inbound or outbound wave of flights is longer.

The point-to-point (P2P) scheduling philosophy is mostly aimed at origin/destination (OD) traffic and is often used on "trunk routes", i.e. those routes with high traffic volumes. Pure P2P networks are widely adopted by LCCs because of the traffic density on these trunk routes and the simplicity of LCC business models (Lawton 2002). P2P networks do not have the desirable "economies of density", but most LCCs still enjoy the "economies of scale" at the level of individual routes and the corporate business level by maintaining a low-cost business base. Since P2P services are aimed at OD markets, pure P2P networks often provide passengers with little or no connections (interlining services), except at those airports serving more destinations in a P2P network, e.g. Luton of easyJet and Dallas-Fort Worth of Southwest.

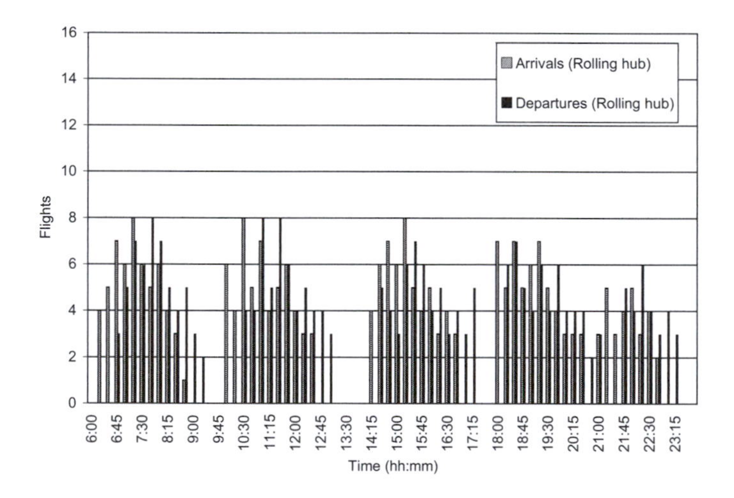

Figure 2.3 A rolling hub schedule (hypothetical data)

The desired attribute of limited connections of pure P2P schedules also limits the growth of this model, especially in two areas: limited inbound traffic from inter-continental operations and limited internal connecting traffic within their own network. Virgin Blue and JetBlue in the U.S. are the best examples where "low-cost interlining agreements" with inbound traffic have been utilised, often from non low-cost carriers or international partners. "Internal" connecting traffic in a P2P network can be facilitated by the "self-connection" of passengers (Malighetti et al. 2008), in which passengers buy two sectors and connect by themselves with no "through check-in or interlining services" provided by the carrier.

2.1.2 Airport Slots and Ground Service Infrastructures at Airports

Regarding the technical considerations in airline scheduling philosophy, these are mostly driven by the requirements to service and maintain aircraft, as well as the physical and operational limits imposed by airports and air traffic control authorities. Among those technical considerations, the issues of landing/take-off slots and ground service infrastructures at airports are the most critical ones.

An airport slot refers to the "right to operate an aircraft (for landing or taking off) at a certain time" (IATA 2006). Slots are allocated to both departure and arrival flights and are limited by the "operating" or "regulated" capacity of an airport (Gilbo 1993). The implication of hubbing schedules is that flights arrive and depart in separate waves with a short period of separation time in between, allowing passenger connections among inbound and outbound flights. The gap between departure and arrival waves is longer than the declared "minimum connection time" at a specific airport. The intensity of arriving and departing flights reflects the strength of hubbing as well as the demand for runway slots. For strong hubbing operations, the demand for slots in the peak of a wave could be more than the capacity, while the demand in the non-peak hours is often significantly lower. The implication for airport capacity is that slot utilisation at a strong hubbing airport is often not efficient.

For rolling hub operations, the difference between peak and trough demand for slots is less than the case of strong hubbing. Accordingly, slot utilisation is more efficient throughout the day, although peaks and troughs may still exist. Flight connection times among high-demand flights are kept short in the case of a rolling hub, while the connection times for other flights could be longer due to the "spread and de-peaked" waves (Goedeking and Sala 2003). For weak hubbing operations, there is usually no clear difference between arrival and departure waves. Accordingly, connections are facilitated among those high-demand flights. For those connections with lower demand, the connection time in a weak hubbing schedule could be longer than the case in a strong hubbing scenario, but under a certain threshold, in order to make the connection commercially competitive in the market.

The high demand for airport slots also increases the demand for ground services and infrastructures at airports. In order to accommodate the arrival flights, airports need to provide enough gates, terminal spaces for connecting passengers, security screening facilities, and baggage/cargo sorting services. The hubbing operation implies that there is high demand for these infrastructures during the peaks of inbound waves and low demand during the off-peak hours. The highly fluctuating demand for airport facilities causes inefficient use of facilities, with low utilisation of facilities during off-peak hours. Since these infrastructures are expensive asset investments for airports, and in some cases, for airlines, low utilisation of airport assets translates into high fixed costs in airline businesses. Under the pressure of cost cutting in the era of post-9/11 terror attack in the U.S. in 2001, and growing competition from LCCs, many network carriers have started depeaking some hubs in the U.S. and in Europe, so as to cut operating costs.

The goal of de-peaking is to utilise more fully airport facilities and reduce "self-induced" delays due to congestion in peak-hour operations. The actual amounts saved moving from strong hubs to rolling hubs are not publicly reported in the industry. It is generally believed that in this scenario airlines lose some connection options, and hence, less products (or less attractive products due to increased connection time) can be offered in the market for connecting traffic. However, benefits are obtained through the higher utilisation of facilities and human resources at airports, as well as strengthening of key connecting markets (Goedeking and Sala 2003). There are significant implications of the scheduling philosophy of airline operations at airports. We have discussed some characteristics of the general scheduling philosophy and the potential impact on airline operations. We will now move on and explore those activities involved in airline operations on the ground at airports.

2.2 Aircraft Turnaround Operations

The goal of ground activities by airlines at airports is to facilitate the transport of passengers and goods under certain safety and security requirements. In addition, the operation of the aircraft itself requires certain engineering services on the ground to prepare an aircraft for a following flight. There are also logistic activities which occur on the ground at airports, providing those resources needed for on-board services such as passenger catering and aircraft fuelling. Aircraft turnaround operations, broadly speaking, cover all of these activities, in which some occur on the landside of an airport, e.g. passenger check-in, and others occur on the airside, e.g. goods loading on the ramp. Since passenger processing and goods handling involve operations on both the landside and airside of an airport, the context of this section covers both airside and landside activities and jointly discusses them from the perspective of airline operations including: aircraft turnaround operations, schedule planning and flight delays.

2.2.1 Passenger Processing

The purpose of passenger processing is to facilitate the movement of passengers at airports, either departing or arriving. Passenger connection between flights involves a slightly different workflow and will be discussed in detail in later sections when we examine flight interlining services. For departing passengers, this process begins with checking in at counters (or on-line check-in as early as 48 hours before the departure time). From the perspective of airline operations, the subsequent processes that are after check-in and before passenger boarding at gates are facilitated by different authorities, including the immigration agency (for international flights), the security screening agency, and the airport authority. This is where airline operations can potentially be disrupted by non-airline processes, contributing to uncertainties in passenger processing.

Departing passengers arrive at boarding gates randomly and wait for the call from ground staff to board the aircraft. Passenger boarding starts from 15 minutes before the departure time for narrow-body aircraft operation such as B737s, to 40 minutes before for jumbo jet operations such as B747s. Depending on airline business models and ground operating procedures, airlines use different boarding methods. For instance, many LCCs adopt the "free-seating" policy for passenger boarding to encourage passengers to arrive at the gate and board the aircraft early, so as to choose preferred seats. This tactic also avoids the process of assigning seats to passengers, so LCCs can reduce operating costs. Other airlines may use the "random" boarding method or even the "back-to-front" method, which is preferred by some network carriers. More passenger boarding methods will be discussed in detail in Chapter 3.

For arriving passengers the processes are simpler, from passenger disembarkation, immigration checks (international flights only), baggage claim, custom checks (international flights only) to leaving the terminal. Little passenger processing by airlines is involved here, except baggage unloading and baggage services for missing bags.

2.2.2 Goods Handling

Goods handling includes the handling of passenger bags and cargo (including mail). Passenger bags are checked in at counters and X-Rayed immediately after passenger check-in before baggage sorting and loading on to an aircraft. A simplified baggage flow at an airport from passenger check in to baggage loading is shown in Figure 2.4. Baggage sorting can be automatic (by RFID or barcodes), semi-automatic or fully manual for small airports, while baggage loading is often operated manually by baggage handlers with loading equipment such as trailers, mini trucks and conveyors. Even in an automatic baggage sorting system, it is often impossible for the scanning and sorting system to reach 100 per cent successful identification. Hence, manual scanning is often involved in an automatic baggage sorting system with a separate manual processing line as illustrated in Figure 2.4.

Depending on the type of an aircraft, baggage can be stored in Unit Load Devices (ULDs), on pallets, or even loosely stored in the belly space for smaller planes. Different ways of baggage processing, storage and loading require different operating times and are subject to different sources of operating disruptions.

For connecting passengers, if a change of plane is involved in a journey, then passenger bags need to be unloaded from the first plane and fed into the baggage sorting system during the connection time of a journey. This process is illustrated in Figure 2.4 in which there is a feeding conveyor that transports inbound connecting bags to the baggage sorting system. Sorted connecting bags are then loaded to the departing aircraft as locally checked-in baggage does. Due to these extra baggage transportation and sorting processes, the connection time between flights at an airport has a specified "minimum time" in order to ensure that connecting bags can be loaded on departure flights before the scheduled departure time. This baggage handling time often significantly influences the "minimum connection time" between flights that an airline can offer at a specific airport. In turn, this constrains airline competitiveness in the commercial world in terms of product offering and the competitiveness of these products. Regarding inbound passenger bags, the process is reversed as shown in Figure 2.5. Custom and quarantine authorities take over baggage screening after passengers have claimed their bags, especially for international inbound travellers.

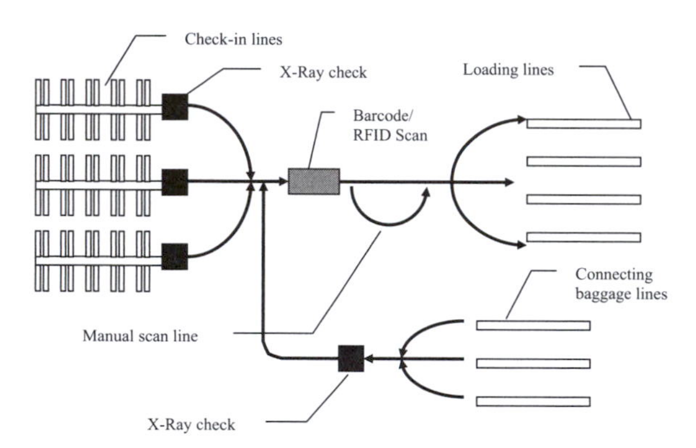

Figure 2.4 Outbound and connecting baggage handling flows

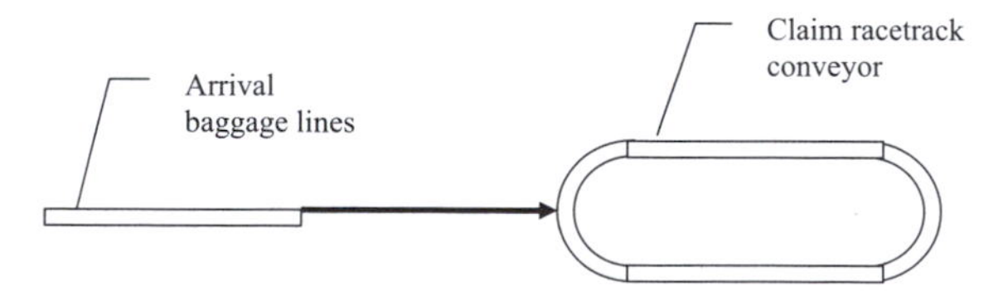

Figure 2.5 Inbound baggage handling flows

Outbound cargo is usually processed at cargo centres close to an airport. Since passenger bags enjoy priority in loading, cargo loading is often processed after the "weight and balance" calculation is finalised for passenger loading. Apart from freighters, passenger aircraft can carry only a certain amount of belly cargo, depending on the available weight of an aircraft after passenger and bags loading, and often more critically, depending on the space available. Belly cargo operation is a relatively small operation when compared with passenger chartering, although it can be significant for some international routes and even more financially significant when the passenger transport volume declines. Belly cargo operations may improve the financial bottom line of airlines during periods of global market downturn such as the one post the burst of the technology bubble and the 9/11 attack in the U.S. in 2001.

2.2.3 Interlining Services

Connecting passengers are aided by "interlining services" between flights by different carriers or with the same carrier. This service simplifies passenger connection via "through check-in" for both passengers and bags between the origin and destination. At the connecting airport where passengers transfer between flights, baggage and cargo are also processed for transfer. The time required to process connecting passengers and baggage at an airport determines the "minimum connection time" of an airline at the airport, which in turn influences the available connections between inbound and outbound flights at hub airports. A shorter minimum connection time between flights creates more connection opportunities within the same connecting window, say 90 minutes. Accordingly, carriers will be able to offer more connections and be more competitive in the connection market via a specific hub. This is why it is so important for major hub airports to maintain efficient passenger and baggage handling in order to maintain the competitiveness of airport operations.

Connecting passengers may go through security screening again before boarding an outbound flight, especially for international flights. Since there is a distance between the arrival gate and the departure gate of an outbound flight (and for international connections, these gates are often located on different levels), connecting passengers need to travel to the departure lounge within a given connection time. This imposes some pressure on connecting passengers, especially when their inbound fight is late.

Connecting baggage and cargo both go through the unloading, sorting and loading processes for outbound flights within the same connection time. Depending on the sorting facilities of an airport, this operation may take less time for newer sorting facilities, e.g. Cathay Pacific at Hong Kong International Airport, and more time for other airports, e.g. two-hour connection from international to domestic flights at some U.S. airports. To ensure passenger and baggage connections, some ground handling agents at hub airports have special operating units whose duty is to ensure urgent connections, with short connection time, are made. For instance, a Quick Transfer Unit (QTU) was developed at Paris Charles de Gaulle (CDG) Airport, helping the hubbing operations of Skyteam, Star Alliance and Oneworld Alliance with up to 3,000 connection possibilities per day in 2006 including connecting bags of passengers (Goldnadel 2007). With the influence of airline alliances increasing, airlines in the same alliance often request to use gates near each other at the same terminal, in order to reduce the connection time for passengers, baggage and cargo. This is already seen at the new Terminal 5 of London Heathrow Airport where Oneworld Alliance will host most operations, and the Terminal 3 of Singapore Changyi Airport, which will be occupied by Star Alliance.

2.2.4 On-board Services

A major part of the "product" that an airline offers to customers is the services that customers receive aboard an aircraft. Regardless of the "classes" of cabins on board, there are two types of services, namely catering services and entertainment services. Catering services provide passengers with snacks, light meals or main meals, depending on the duration of the flight, and the time within the flight the service is offered. Now passengers on some domestic flights are not provided with meals but only snacks and drinks. This is partly due to cost cutting by airlines and partly due to the realisation that not all passengers require meal services, especially for a short travel time aboard. For those passengers who need catering services, some airlines have successfully converted this service from a "cost" item to the airline business in the past to a "revenue" item by offering on-demand catering services on a user-pay basis. This is the tactic adopted by most low-cost carriers and gradually by full-service carriers around the world. On mid- to long-haul operations, this strategy has also been adopted by low-cost carriers, e.g. Jetstar Airways in Australia.

In aircraft turnaround operations, catering services are conducted independently with other turnaround activities, depending on the location of galleys on an aircraft. For narrow-body jets, galleys are often located at the rear of an aircraft, so catering unloading and loading can take place at the same time as passenger disembarkation and boarding. For larger wide-body jets, passenger disembarkation and boarding interfere with catering processing at the galleys aboard the plane, so there are usually separate catering workflows to facilitate passenger boarding, catering loading and cabin cleaning activities during aircraft turnaround operations. The reduction in complementary catering services on board also reduces the time required to finish catering unloading and loading; accordingly, this reduces the time required for aircraft turnaround operations. This has a profound impact on designing the procedures of aircraft turnaround operations and on increasing aircraft utilisation, due to less time spent on the ground for aircraft. More details on aircraft turnaround procedures will be examined in Chapter 3.

Entertainment services include newspapers, magazines and in-flight entertainment programmes. With the increased offering of on-demand in-flight entertainment programmes, airlines need to prepare these programmes during aircraft turnaround time, especially for long-haul international flights. Some programmes are pre-loaded on entertainment systems aboard an aircraft, while some programmes are uploaded for specific flights during turnaround. The uploading of video/audio programmes can take time, depending on the channel of uploading (wireless or wired) and the media used for loading. Live television programmes have gained popularity since JetBlue Airways introduced its DirectTV service across its fleets back in 2000 (Shifrin 2004). The DirectTV service utilises satellite communication channels, so avoids the need to upload entertainment programmes during aircraft turnaround.

2.2.5 Aircraft Services

A major duty in the aircraft turnaround process is providing the required services to the aircraft itself such as: refuelling, routine visual engineering check, and auxiliary power unit (APU) services. Refuelling an aircraft is perhaps the most critical activity (service time-wise) for LCC short-haul operations, due to the very limited turnaround time for each flight. Some flights by Southwest, Ryanair and easyJet are reported to have a turnaround time for B737 as short as 15 minutes. Even a longer 20-minute turnaround time can put some pressure on aircraft refuelling, depending on whether a fuel truck or only a pump truck is needed on the ramp of an airport. The use of fuel trucks to transport fuels from depots to aircraft gates takes time and can be disrupted due to the logistic scheduling of fuel trucks on the tarmac.

Routine engineering checks are also conducted on the ramp and usually involves visually inspecting the aircraft body and engines. This service is to ensure that there is no visible damage to engines and aircraft body due to "foreign objects" (e.g. bird strikes), and the aircraft is suitable to conduct those following flights scheduled for the remaining day. In some scenarios, further inspection will be conducted beyond visual inspection if the monitoring systems on an aircraft indicate such a need. If prompt engineering service cannot resolve the issue, further aircraft down time is needed and may cause long delays and schedule disruption. Auxiliary Power Units (APUs) are used to provide an aircraft with external power when engines are switched off. This is usually the case when air conditioning is required for an aircraft, if the aircraft is parked at a remote airport stand for an extended period of time in cold or hot conditions. However, the use of APUs on the apron often causes air pollution issues because of the fuel burning by some APUs.

The activities involved in aircraft turnaround operations are best described by Figure 2.6. The figure shows the activities required to turn around a B737 for domestic operations by a carrier. There are two important messages to be taken from this chart. First, some activities are conducted sequentially on the time line, e.g. cabin cleaning starts after passenger disembarkation. Accordingly, delays to some services may cause delays to other services "down the line". Second, the service time of activities determines the total required time for turning around an aircraft, meaning the shorter the individual service time of each activities, the shorter the total aircraft turnaround time is. This explains how LCCs can pursue short turnaround time and high daily aircraft utilisation. This chart will also play an important role when we discuss the allocation of aircraft ground time in a schedule in the next section and the modelling of aircraft turnaround operations in Chapter 3.

2.3 Schedule Constraints and Delays

2.3.1 Delays and Schedule Flexibility

Schedule delays are common occurrences in airline operations, given the many operational tasks involved, the stochastic nature of operating time, and unexpected disruptions in operations. In addition, airlines operate in an environment in which airlines have limited control over the system constraints, including airport capacities and airspace capacities. The complex interaction between the planned (and fixed) airline schedules and stochastic disruptions may cause flight delays.

Flight delay refers to the time difference between the scheduled departure/arrival time and the actual departure/arrival time of a flight on the day of operation. By definition, it may occur that delays are "negative", meaning early departure or early arrival of a flight. Negative delays are often not an issue and occur when the schedule is running close to plans. However, early departures and arrivals can cause minor issues for airport operations, because early departure requests may disrupt the sequencing of departing flights and early arrivals may disrupt the allocation of gates, especially during peak hours at busy airports.

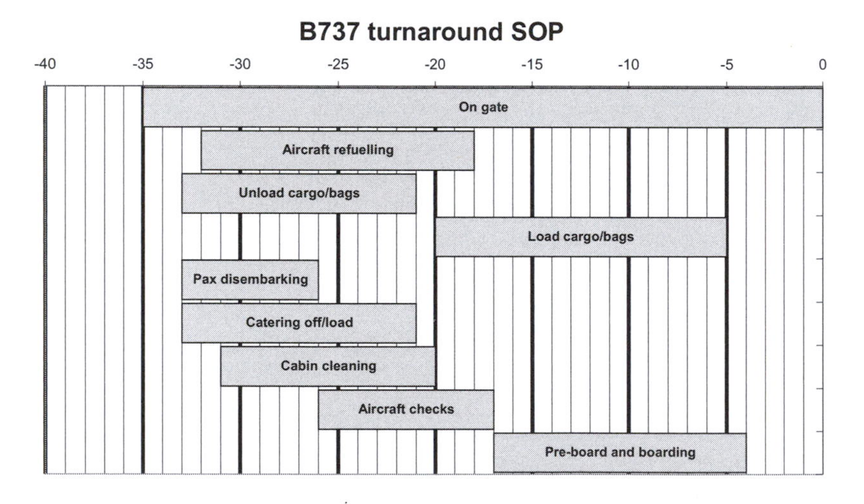

Figure 2.6 B737 turnaround operations for domestic flights

On the other hand, "positive" flight delays are often causes of concern for airlines and passengers. Flight delays are frequently cited by the industry and aviation research to be among the important factors which may significantly impact passenger satisfaction and re-purchase intention in the future (Doganis 2002; Holloway 2008), and even the market share and financial performance of an airline (Bhat 1995; Suzuki 2000). Analytically, the relationship between the planned schedule, arrival (inbound) delay, operational delay and departure (outbound) delay can be modelled by Figure 2.7. For the convenience of delay modelling in the remaining sections of this chapter, the notations used hereafter (including those in Figure 2.7) are summarised below and are also seen in the Appendix of this chapter.

f_{ij}	flight *i* of route *j* that departs Airport A and arrives at Airport B (as in Figure 2.7)
$f_{(i-1,j)}$	the flight flown before f_{ij} on route *j* operated by the same aircraft
s_{ij}^{A}	the scheduled time of arrival of f_{ij}
t_{ij}^{A}	the actual time of arrival of f_{ij}
$s_{(i-1,j)}^{A}$	the scheduled time of arrival of $f_{(i-1,j)}$
$t_{(i-1,j)}^{A}$	the actual time of arrival of $f_{(i-1,j)}$
s_{ij}^{D}	the scheduled time of departure of f_{ij}
t_{ij}^{D}	the actual time of departure of f_{ij}
S_{ij}^{TR}	the scheduled turnaround time of f_{ij} at Airport A
S_{ij}^{BX}	the scheduled block time of f_{ij}
$d_{(i-1,j)}^{A}$	the arrival delay of $f_{(i-1,j)}$ and $d_{(i-1,j)}^{A} = t_{(i-1,j)}^{A} - s_{(i-1,j)}^{A}$
d_{ij}^{D}	the departure delay of f_{ij} and $d_{ij}^{D} = t_{ij}^{D} - s_{ij}^{D}$
d_{ij}^{A}	the arrival delay of f_{ij} and $d_{ij}^{A} = t_{ij}^{A} - s_{ij}^{A}$
d_{ij}^{OP}	the delays due to turnaround operations of f_{ij}
${}_{G}S_{ij}^{b}$	the scheduled ground buffer time of f_{ij} at Airport A
${}_{A}S_{ij}^{b}$	the scheduled airborne buffer time of f_{ij} en-route Airport A and B

$\hat{h}_i$	the realised (actual) turnaround time of f_{ij}
$f_i(h)$	the stochastic distribution of $\hat{h}_i$ with mean value, $\overline{h}_i$
$\hat{k}_i$	the realised (actual) flight time of f_{ij} between Airport A and B
$f_i(k)$	the stochastic distribution of $\hat{k}_i$ with mean value, $\overline{k}_i$
$f_i(t)$	the stochastic distribution of the actual arrival time of f_{ij}, i.e. t_{ij}^A
$g_i(t)$	the stochastic distribution of the actual departure time of f_{ij}, i.e. t_{ij}^D
$g_i'(d_{ij}^D)$	the function of departure delay; $g_i'(d_{ij}^D) = g_i(t_{ij}^D - s_{ij}^D) = g_i(t) - s_{ij}^D$
m_1	the efficiency of delay absorption by the scheduled buffer time of f_{ij}
m_2	the efficiency of turnaround operations at Airport A
$C_P(d_{ij}^D)$	the passenger delay cost function with a marginal delay cost function, $\gamma_P^m(d_{ij}^D)$; a function of delay time d_{ij}^D
$C_{AC}(d_{ij}^D)$	the aircraft delay cost function with a marginal delay cost function, $\varphi_{AC}^m(d_{ij}^D)$; a function of delay time d_{ij}^D
$C_{AL}({}_G S_{ij}^b)$	the opportunity cost of aircraft time with a marginal schedule time cost function, $\delta_{AL}^m({}_G S_{ij}^b)$; a function of ground schedule buffer time, ${}_G S_{ij}^b$
C_T	the total cost of the schedule time optimisation model, including the cost of delays, D_C and the cost of schedule time, S_C
D_C	the expected cost of delays including passenger delay cost, $C_P(d_{ij}^D)$ and aircraft delay cost, $C_{AC}(d_{ij}^D)$
S_C	the cost of schedule time calculated by $C_{AL}({}_G S_{ij}^b)$
α	the weight factor, representing the trade-off between delay cost and schedule time cost

In Figure 2.7, flight f_{ij} originates from Airport A and is scheduled to leave for Airport B by the scheduled time of departure, s_{ij}^{D}. $f_{(i-1,j)}$ is the flight flown before f_{ij} on "route" j that is operated by the same aircraft. A *route* is a series of flight legs that is assigned for operation by an aircraft within a period of time. For illustration purposes, we assume that flight $f_{(i-1,j)}$ arrives late at Airport A by an arrival delay of $d^{A}_{(i-1,j)}$ which is analytically defined by $d^{A}_{(i-1,j)} = t^{A}_{(i-1,j)} - s^{A}_{(i-1,j)}$. Since flight $f_{(i-1,j)}$ and f_{ij} are operated by the same aircraft on the same route j, flight f_{ij} incurs an inbound delay of the same amount, $d^{A}_{(i-1,j)}$ as shown in Figure 2.7.

The common methodology to design the scheduled turnaround time (S_{ij}^{TR}) for f_{ij} is to combine two components in the turnaround time, namely the mean turnaround time to finish ground services and the ground buffer time, ${}_{G}S_{ij}^{b}$ to compensate arrival delays due to late inbound aircraft, i.e. $d^{A}_{(i-1,j)}$ or delays occurred in turnaround operations, d_{ij}^{OP}. In some cases, the scheduled turnaround time also needs to consider the time required for passenger connections from other inbound flights to f_{ij}, especially if a hubbing schedule is operated at Airport A.

If delays (d_{ij}^{OP}) due to disrupting events in turning around flight f_{ij} at Airport A together with the inbound delay $d^{A}_{(i-1,j)}$ are higher than the scheduled ground buffer time ${}_{G}S_{ij}^{b}$, then flight f_{ij} may incur departure (outbound) delay by d_{ij}^{D}, depending on the "realised" turnaround time of f_{ij} on the day of operation. d_{ij}^{D} is analytically defined by $d_{ij}^{D} = t_{ij}^{D} - s_{ij}^{D}$. Disrupting events in this context may come from aircraft turnaround operations as well as from connecting crew and passengers from other flights.

A certain amount of airborne buffer time, ${}_{A}S_{ij}^{b}$ is often embedded in the block time of flight f_{ij} to absorb delays due to airport congestion at both ends of the flight. If flight f_{ij} is further delayed en-route and this delay together with the outbound

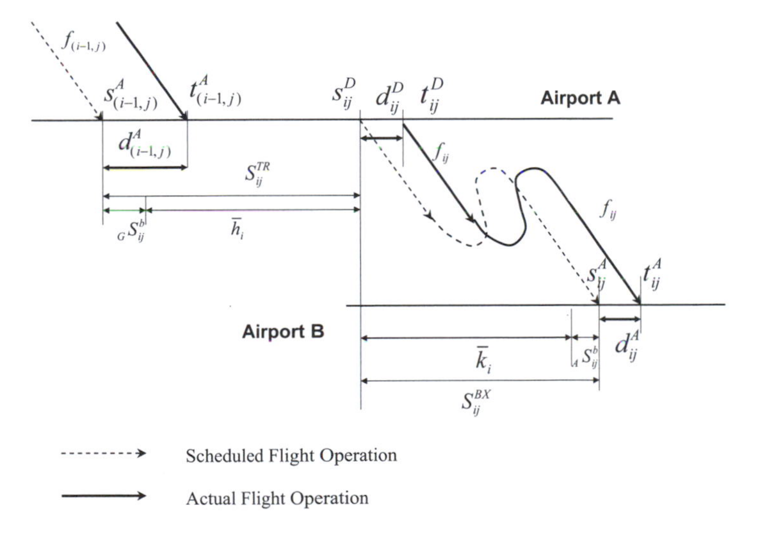

Figure 2.7 The relationship between schedules, turnaround times and delays

delay, d_{ij}^{D} cannot be fully compensated for by the airborne buffer time, ${}_{A}S_{ij}^{b}$, then flight f_{ij} would incur arrival (inbound) delay at the destination Airport B, i.e. d_{ij}^{A} ($d_{ij}^{A} = t_{ij}^{A} - s_{ij}^{A}$). This example also demonstrates how inbound delays and ground operational delays to flight f_{ij} at Airport A may cause the arrival delay of flight f_{ij} at Airport B.

Delays due to disruptions of ground operations at Airport A may exacerbate the situation by cancelling out the function of the built-in ground buffer time of f_{ij}. The inbound delay at Airport B may further cause delays to other flights that receive inbound connecting passengers, crew or goods from f_{ij} at Airport B, causing delay propagation within the network. The development of delays within a network will be further discussed in Chapter 4, when the concept of network operation is introduced.

2.3.2 Stochastic Service Time of Turnaround Activities

Airline schedules and flight timetables are usually planned well ahead of the operation. The challenge of managing aircraft turnaround operations is to ensure that ground operations for turning around an aircraft finish within the planned turnaround time, as denoted by S_{ij}^{TR} in Figure 2.7 earlier. The crux of this challenge is in the fact that the service time for most activities in turnaround operations is stochastic in nature. Passenger processing time is often proportional to the number of boarding and connecting passengers of a specific flight. Accordingly, some on-board services for passengers take longer to accomplish, when more passengers are boarding an aircraft, e.g. catering services and baggage processing. Aircraft re-fuelling may also take longer, as the take-off weight of an aircraft increases.

Due to the stochastic nature of operating times of ground services, one way to alleviate the impact of this uncertainty is to incorporate a buffer time in the scheduled turnaround time, e.g. the ${}_{G}S_{ij}^{b}$ in Figure 2.7. Hence, the actual (realised) turnaround time of f_{ij} on the day of operation determines departure delays and can be modelled by (2.1):

$$d_{ij}^{D} = d_{(i-1,j)}^{A} + \hat{h}_{i} - S_{ij}^{TR} = d_{(i-1,j)}^{A} + \hat{h}_{i} - ({}_{G}S_{ij}^{b} + \bar{h}_{i}) \tag{2.1}$$

where $\hat{h}_{i}$ is the realised turnaround time of f_{ij} and is a stochastic variable; the distribution of $\hat{h}_{i}$ is $f_{i}(h)$ and is a stochastic function with a mean value of $\bar{h}_{i}$, i.e. the "mean turnaround time" as shown in Figure 2.7. It can be seen from (2.1) that airline schedules form a "fixed boundary", i.e. S_{ij}^{TR}, while other variables in (2.1) are stochastic, causing the departure delay of f_{ij}, d_{ij}^{D} a stochastic variable as well. The flexibility of airline schedule comes from the planned ground buffer time, i.e. ${}_{G}S_{ij}^{b}$ in (2.1), that is designed to compensate inbound delays to f_{ij}, $d_{(i-1,j)}^{A}$. There are chances that S_{ij}^{TR} may over-compensate the total of inbound delay and actual turnaround time, $(d_{(i-1,j)}^{A} + \hat{h}_{i})$. For these cases, an early departure with a "negative" departure delay is possible.

2.3.3 Attributes of Turnaround Operations

Since the service times of activities in aircraft turnaround operations are stochastic and may depend on passenger load, the time required to turn around an aircraft is stochastic as well. As discussed briefly in Section 2.2 earlier, there are two important attributes of airline ground operations. First, some activities need to be conducted sequentially on the timeline, while some can be conducted on parallel with other activities. The implication is that any disruptions to a component of ground operations can cause "knock-on" effects to other processes and can cause delays to departure flights. For instance, a delay of baggage unloading at the apron may cause a delay to baggage loading, which may also cause delays to connecting passengers and goods to other flights at the same airport.

Second, the total time required to turn around an aircraft has a profound impact on airline scheduling and airline operations. When various aircraft turnaround activities are treated aggregately as a "single" process, we will not observe the influence of individual processes on the whole operation. Instead, the turnaround operation is seen as a "black box" which receives an input delay (from an inbound flight) and generates an output delay (for an outbound flight). This approach provides us with a "macro" view of aircraft turnaround operations and simplifies the observation and modelling work needed to study airline schedules. Also, this approach provides us with the "standard" turnaround time needed for different types of fleets and sometimes for similar operations at different airports in a network.

With historical data of flight operations, one can plot the distribution curve, i.e. the probability density function (PDF) of the "realised" turnaround time, $f_i(h)$ for flight f_{ij} over a period of time. These PDFs can reveal important information such as: (a) the mean departure and arrival delay of f_{ij}, i.e. d_{ij}^D and d_{ij}^A; and (b) the variance of departure and arrival time of f_{ij}, i.e. the range of t_{ij}^D and t_{ij}^A. Together with data of the actual turnaround time ($\hat{h}_i$), the PDFs of flight delays help airlines conduct post-operation delay analyses and design the optimal turnaround time for ground operations. Accordingly, this helps shape airline schedules, networks and aircraft utilisation and improve operational efficiency as well as financial bottom line.

2.4 The Optimal Turnaround Time – Empirical Methods

2.4.1 The Pursuit of Optimal Turnaround Time

It can be seen from the discussions in previous sections that airline schedules impose constraints on airline operations, both on the ground and en-route between airports. Hence the design of an airline schedule is critical in that the desired or the ideal schedule should provide airlines with adequate "flexibility" to absorb unexpected disruptions and delays, while the schedule should also generate high "productivity" of airline assets by utilising aircraft time for revenue-making block time. Accordingly, airlines design aircraft turnaround time to absorb inbound delays due to late aircraft. The aircraft turnaround time is designed to minimise the impact of disruptions on flight departures during airline ground operations. This is a trade-off situation in which on one hand, long turnaround time reduces delays and stabilises airline operations, but on the other, long turnaround time reduces the utilisation of aircraft, because ground time could otherwise be used somewhere else in the network as revenue-making block time. Hence, there is always a desire to pursue the optimal turnaround time design for airlines.

Two models will be introduced in this section and the next. The first model is an ad hoc approach, namely the *Empirical Model* that involves the use of statistical techniques and stochastic theories to model distributions of flight delays and determine the optimal allocation of turnaround time and block time for a schedule. This approach provides a quick solution to the issue of turnaround/block time allocation by considering the uncertainties involved in real operations. The second model is an aggregate analytical model, namely the *Turnaround Time Allocation* (TTA) model, which involves modelling the stochastic distributions of flight delays and delay costs in order to explore the trade-off scenarios during airline schedule planning. This approach provides an analytical tool for exploring and evaluating scheduling policies, which is often required at the early stages of strategic schedule planning. Both models are based on stochastic and statistical theories, reflecting the stochastic nature of airline operations and the uncertainties involved in the environment of airline operations.

2.4.2 The Empirical Model

This model is widely used by airlines as an ad hoc approach to model flight delays with uncertainties. The theoretical foundation of this model is based on stochastic theories, by which the actual arrival time of a flight is modelled as a stochastic variable, denoted by t_{ij}^{A}. The collection of previous actual arrival times of the same flight (f_{ij}) over a period forms a stochastic distribution, denoted by $f_i(t)$. The actual departure time of f_{ij} is also a stochastic variable, denoted by t_{ij}^{D} and the distribution of departure time is $g_i(t)$.

Historical data are used to plot the probability density function (PDF) of a chosen flight. According to stochastic theory, the more samples one has, the more closely the PDF will resemble the real curve, which is unknown. When the arrival time samples are plotted against the frequency of occurrence, a PDF may look like Figure 2.8. The PDF of the example Flight 902 is represented by the line with box legends. Since early arrivals (with "negative" delays) are treated as zero delays, the PDF function does not span to the negative side of the x-axis. The purpose of plotting the PDF curve is to visually study the delay pattern of this flight.

PDF curves can be converted into cumulative density functions (CDFs) as the line shown in Figure 2.8 with triangle legends. With CDF curves, we can clearly see the on-time performance (OTP) of the study flight according to past operations.

The on-time probability (zero delay, denoted by "D0") is around 40 per cent, while the 15-minute delay probability (denoted by "D15") is about 70 per cent as shown by the arrows on the figure. Similar methods can be used to construct the PDFs and CDFs when studying departure patterns of flights.

For some situations, it is a good idea to model aircraft turnaround processes aggregately. One reason is to simplify the modelling process as well as the model itself. The other reason is often due to the limited availability of operating data, or modelling "convenience" at the time of model building. In our Empirical Model, the aircraft turnaround process is modelled as a "black box" process, i.e. aggregately as a "single" process combining all activities. Hence, the relationship between the arrival time PDF of an inbound flight $f_{i-1}(t)$, turnaround operation of f_{ij} and the departure time PDF of an outbound flight $g_i(t)$ can be described by Figure 2.9.

Analytically, the distribution $g_i(t)$ is a function of three variables as formulated previously in (2.1), including the actual time of arrival $t^A_{(i-1,j)}$, the actual turnaround

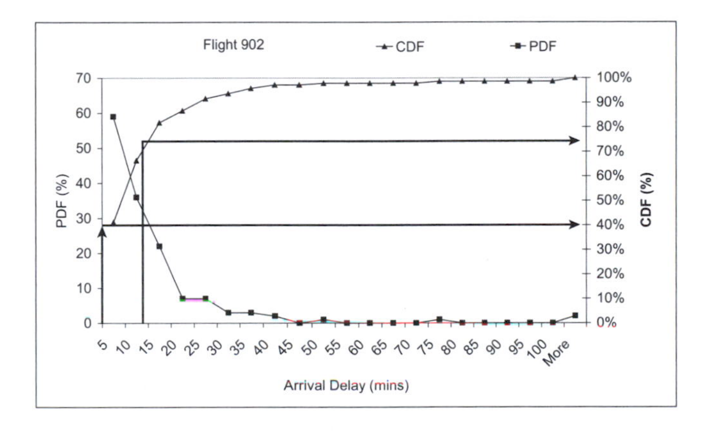

Figure 2.8 The PDF and CDF of Flight 902

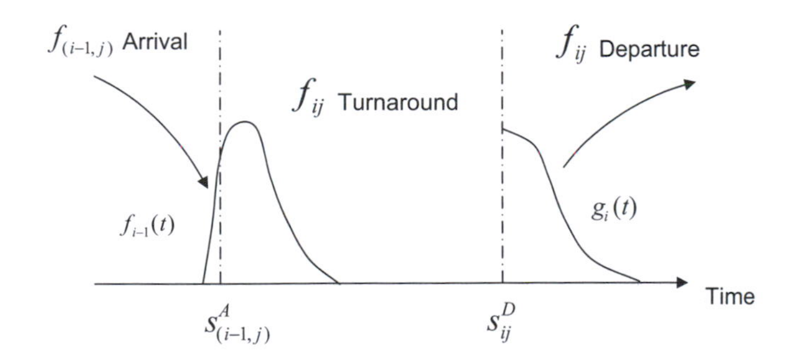

Figure 2.9 Relationships between arrival PDF and departure PDF

time $\hat{h}_i$, and the scheduled turnaround time S_{ij}^{TR}. The objective of the following model is to minimise the probability of incurring delays by adjusting the scheduled turnaround time, S_{ij}^{TR}. Since the actual turnaround time $\hat{h}_i$ is stochastic, the objective is to design a turnaround time that is long enough to allow $\hat{h}_i$ and meanwhile, allow some buffer to cover inbound delays and operational delays on the ground.

2.4.3 Model Implementation

A simple and straightforward approach to implement the Empirical Model is to plot the CDF of the actual turnaround time ($\hat{h}_i$) from the PDF of $\hat{h}_i$, i.e. $f_i(h)$. The resulting CDF, $F_i(h)$ is illustrated in Figure 2.10. In this example, the scheduled arrival time of the inbound flight before Flight 208 is at 17:50 and the scheduled departure time at 18:20, allowing a turnaround time of 30 minutes for Flight 208. The standard turnaround time for the specific aircraft type (B737) is assumed to be 25 minutes by a low-cost carrier (hence, five minutes buffer time). The actual turnaround time CDF of Flight 208, $F_i(h)$ in Figure 2.10 shows that 50 per cent of past turnarounds took less than 30 minutes.

Airlines often have an operational target, say 80 per cent for turnaround operations, which is the on-time probability target that airlines wish to "cover" by the scheduled turnaround time. In other words, according to historical data of the study flight, there will be an 80 per cent chance that the actual turnaround time of a real operation takes less than the designed turnaround time, which is 37 minutes in this example as shown in Figure 2.10 above (the mean of the bin 35–40 mins is taken as the mid point of the chosen bin on the x-axis). Analytically, the difference between 37 and 25 minutes is the required ground buffer time, i.e. 12 minutes to achieve the 80 per cent coverage target. Accordingly, the optimal scheduled departure time for Flight 208 becomes 18:27.

The previous approach did not consider the influence of inbound delays on departure delays of Flight 208. Ideally, the average inbound delay of $f_{(i-1,j)}$ would be minimum, because most inbound delays of $f_{(i-1,j)}$ are expected to be absorbed by the scheduled block time of the inbound flight. In reality, the average inbound delay of $f_{(i-1,j)}$ is often significant, although not always large. In the above example, if the average inbound delay of $f_{(i-1,j)}$ is $\bar{d}_{(i-1,j)}^{A}$, then the CDF plotted for allocating turnaround time should be the actual departure time of f_{ij}, i.e. $g_i(t)$ from historical data and the CDF of $g_i(t)$, noted by $G_i(t)$ as seen in Figure 2.11. To achieve 80 per cent departure punctuality, the optimal scheduled departure time for Flight 208 in this example will be 18:37 (taking the mid point of the category bin between 18:35 and 18:40). Accordingly, the scheduled turnaround time becomes 47 minutes, which is higher than the previous result (37 minutes) that did not consider regular inbound delays.

The second approach described here is used by airlines more often than the previous one, because of the significant impact of inbound delays for most flights.

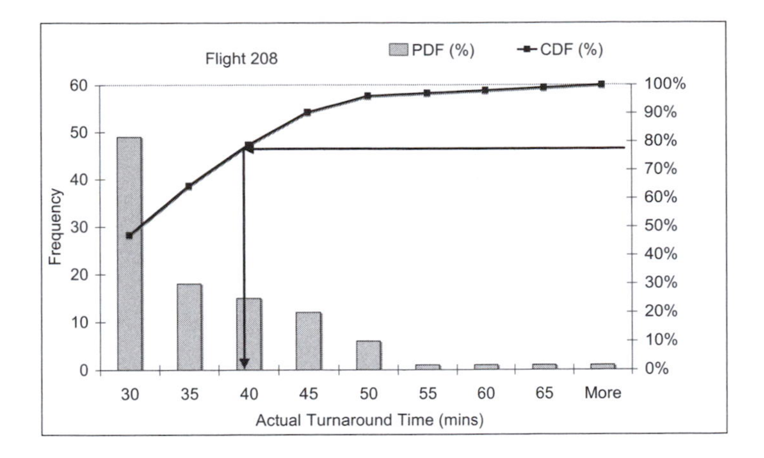

Figure 2.10 The PDF/CDF of actual turnaround time of Flight 208 samples

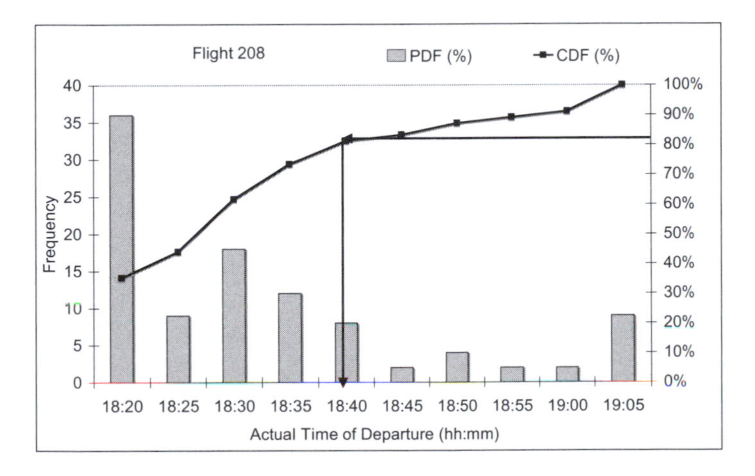

Figure 2.11 The PDF/CDF of actual departure times of Flight 208

It is clearly seen from this example that by considering inbound delays due to late aircraft arrival, the low-cost carrier in this example needs to schedule 47 minutes for Flight 208, in order to achieve the schedule reliability target, i.e. 80 per cent. Hence, this approach results in scheduling more time for aircraft turnarounds. One can also see that if the carrier focuses on reducing inbound delays, then the required turnaround time for this example can be reduced from 47 minutes to 37 minutes, thus leading to an improvement in aircraft utilisation and flight punctuality.

This approach is a common scheduling technique used by LCCs as well as network carriers. However, the delay propagation effect on other flights in a network is not always clear to airlines. Hence, most airline schedules do not necessarily reflect this consideration. Without this consideration in schedule planning, one can observe clear delay propagation in schedule operations, especially for LCCs and strong hubbing airlines. To counter this potential impact on schedule reliability, LCCs often deploy long buffer times in the midday as a "fire break time" in order to control delay propagation in a network. In addition, the choice of using secondary airports by LCCs also reduces the risk of incurring operational delays on the ground, because these airports are usually regional airports and have less traffic (and congestion) than major airports. In turn, this reduces the likelihood of delay propagation and its impact on airline operations.

2.5 The Optimal Turnaround Time – Analytical Methods

2.5.1 Analytical Models

The Empirical Model introduced in Section 2.4 is often used as a quick approach to evaluate current flight delay patterns and how the planned schedule is functioning, especially on the adequacy of scheduled ground time in the schedule. The Empirical Model, however does not allow an analyst to evaluate scheduling questions such as the variation of scheduled turnaround time and the efficiency of turnaround operations at different airports. In order to evaluate the potential impact of schedule changes on flight delays and delay related costs, an analytical model, called the *Turnaround Time Allocation* (TTA) optimisation model is introduced in this section.

The objective of this model is to provide airline schedulers with an evaluation tool that can be applied at the stage of schedule planning for different scheduling considerations. This is especially valuable for considering trade-offs involved in airline schedule planning such as the use of buffer time and the efficiency of ground operations. Some simplifications and assumptions are made during the following modelling processes. Although simplification in modelling may cause gaps between the model and the real system, simplification is commonly seen in analytical approaches due to limited resources such as time and budget, and potentially numerous constraints in problem solving (Klafehn et al. 1996).

2.5.2 Delay Development Mechanism

The relationship between departure delay, arrival delay of the previous flight (by the same aircraft), aircraft turnaround operation time and the scheduled turnaround time in a schedule is previously formulated by (2.1). We can see that aggregately, the departure delay of a flight is influenced by the inbound arrival delay of the aircraft as well as the actual time required to finish aircraft turnaround operations. The scheduled turnaround time is designed to control delays from both the inbound aircraft and ground operations. If the turnaround process of an aircraft is modelled as an "input-output model", then the "input" is arrival delay and the "output" would be departure delay. A critical component in an "input-output model" is the description of the mechanism by which an input is converted or transformed into an output. The transforming mechanism in modelling flight delays is the process by which inbound delay develops into outbound delay by considering the effects of the scheduled turnaround time, i.e. S_{ij}^{TR}, including the planned ground buffer time, ${}_{G}S_{ij}^{b}$.

The scheduled ground buffer time is designed to absorb arrival delays and unexpected delays due to ground handling disruptions. The mean turnaround time, $\bar{h}_i$ represents the standard service time for ground handling agents to complete operational procedures to turn around an aircraft for a following flight. Due to the complexity of aircraft turnaround procedures, delays to turnaround aircraft may be caused by many factors such as ground handling equipment serviceability, passenger connections, passenger delay, and aircraft arrival delay. A critical function of turnaround operations by airlines, apart from preparing aircraft for subsequent flights, is to decrease the magnitude of departure delay to the lowest possible level within the given turnaround time. Hence, the operational efficiency of ground operations is critical in controlling delay propagation in an airline network, due to flight connections by aircraft routing.

The development mechanism of departure delay is illustrated in Figure 2.12. If the arrival delay of $f_{(i-1,j)}$, noted by $d_{(i-1,j)}^{A}$ is shorter than the scheduled buffer time, ${}_{G}S_{ij}^{b}$ for the turnaround of f_{ij}, then the arrival delay will be partially or fully absorbed by the scheduled buffer time. The delay absorption "performance" of schedule buffer time is denoted by m_I, i.e. the slope of the early portion of the delay time development curve before ${}_{G}S_{ij}^{b}$ as shown in Figure 2.12. When the arrival delay is larger than the buffer time ${}_{G}S_{ij}^{b}$, the resulting departure delay of f_{ij} may develop according to one of three scenarios. First, as indicated by curve f_2 in Figure 2.12, departure delay may develop in a linear form proportional to the amount of arrival delay, no matter how long the arrival delay is.

Second, ground handling agents may be able to maintain efficient turnaround operations under the pressure of limited turnaround time and buffer time. In such a scenario, the level of departure delay will be less than the one of inbound delay and will follow curve f_3. This is a desired scenario in which the planned turnaround time by airlines functions effectively as delay buffer and airlines are able to contain the scale of delay propagation in a network without tactically allocating more resources during operations.

Third, curve f_1 represents a typical situation in which ground operations are further disturbed by the late arrival of inbound aircraft, via late transfer of passengers, late passenger check-in, late baggage connection, or disruptions in ground handling plans. One major responsibility of airline dispatchers (or called ground controllers or ground co-ordinators) at airports is to deliver punctual departures by "operational means". Ways in which this can be done may include: skipping the loading of some cargo or goods that are not urgent (and can be deferred to a later flight), or allocating more resources to speed up aircraft turnaround. This type of "speeding up" of operations is commonly seen in the industry by both

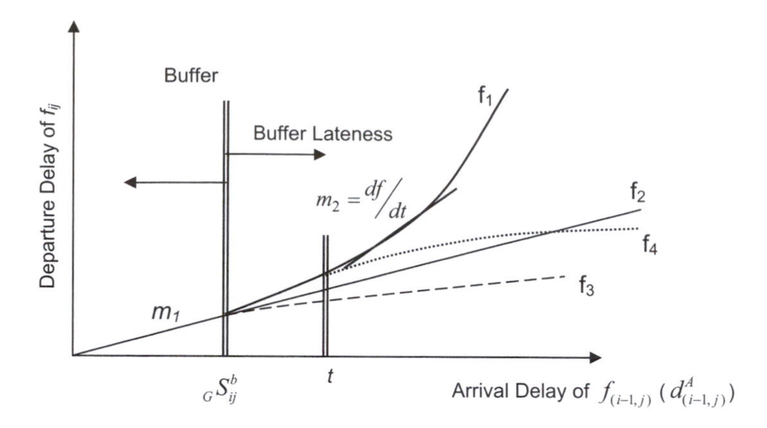

Figure 2.12 Development of departure delay due to arrival delay of inbound aircraft

network carriers and LCCs and is often used as a means of "fire breaking", in order to stop or significantly reduce the risk of delay propagation in an airline network.

If at time t (as shown in Figure 2.12), the airline dispatcher takes actions to reduce departure delay of flight f_{ij}, the development of departure delay following curve f_1 might become curve f_4 (illustrated by the dotted line). Accordingly, f_{ij} may incur a shorter departure delay than expected in the "no-action" scenario, i.e. curve f_1. In addition, the reduction of the departure delay to f_{ij} may reduce the risk of incurring potential knock-on delays through aircraft rotations and flight connections in a network. Nevertheless, the operating cost of f_{ij} may increase in this scenario.

The curve slope (denoted by m_2 and shown in Figure 2.12) after the turnaround buffer time (${}_G S_{ij}^b$) is used to define the "efficiency" of ground services, i.e. the ground handling agents' capacity to respond to schedule perturbations. When the value of m_2 is less than or equal to one, departure delay develops at a lower rate than arrival delay such as curve f_2 and f_3. If m_2 is greater than one, it means that turnaround operations are further disturbed by operational disruptions or inadequacy of resources and hence, ground operations take a longer time to complete. Consequently, outbound departure flights will suffer further delays, not only due to arrival lateness but also from operational disturbance during aircraft turnaround.

The "efficiency and performance" issue of ground handling has been a critical one in airline ground operations due to the continuing pressure of cost cutting in the airline industry. To ensure the efficiency of ground operations and maintain the capacity to absorb delays during turnaround time, airlines often establish service level agreements (SLAs) with ground handling agents to achieve the goal, especially if ground handling services are outsourced to a third-party handler. SLAs outline the "benchmarks" of aircraft turnaround operations by a number of indices such as average delay time, and the occurrence of delay causes. However, most SLAs between airlines and ground handlers are commercially confidential and not available in the public domain.

2.5.3 Delay Development Modelling

An analytical aircraft turnaround model is developed based on the rationale of the delay development mechanism described previously. The departure time of flight f_{ij} is previously modelled by (2.1) as a function of the arrival time of previous flight, $f_{(i-1,j)}$ and the turnaround time S_{ij}^{TR} of f_{ij}. From observations of airline operations at airports, it is seen that whenever there is little or no buffer time (i.e. ${}_{G}S_{ij}^{b}$ is small), departure delay can be modelled by *Curve A* as illustrated in Figure 2.13. When the scheduled buffer time is as long as the maximum limit, i.e. $\max({}_{G}S_{ij}^{b})$, departure delay can be modelled by *Curve C* and nearly 100 per cent arrival delays are covered.

Often, airlines are not willing to schedule excessive buffer times. However, when adequate buffer times are scheduled, any arrival delay can be absorbed by the buffer time and it is likely that there will be very little departure delay. For delays longer than the $\max({}_{G}S_{ij}^{b})$, it is assumed that delay will develop as *Curve A*. In between these two extreme cases, departure delay can be modelled by *Curve B* with a scheduled turnaround time (S_{ij}^{TR}) including a buffer (${}_{G}S_{ij}^{b}$). This is a one-to-one mapping function between the arrival time and departure time of two flights that are operated sequentially by the same aircraft. For any given buffer time ${}_{G}S_{ij}^{b}$, there will be a corresponding "efficiency" curve as illustrated by *Curve B* in Figure 2.13, which represents the operational efficiency of ground services under the constraint of scheduled turnaround time, S_{ij}^{TR}.

The "mapping function" (*Curve B*) is modelled as a piece-wise linear function in order to describe one of those potential scenarios of delay development shown previously in Figure 2.13. Other function forms besides linear functions can be adopted. Although complex function forms would allow analysts more flexibility and the capacity to address complex delay development mechanism, complex function forms may not be solvable analytically and would require advanced techniques to solve the resulting model.

The "efficiency" of delay absorption of schedule buffer time ${}_{G}S_{ij}^{b}$ is modelled by m_1, which is the slope of the left portion of *Curve B*. The operational efficiency of aircraft turnaround operations in dealing with delays and resource constraints is modelled by m_2, which is the slope of the right portion of *Curve B*. The parameter value of m_2 is assumed to be given from statistical analysis and is airport and handler specific, reflecting the unique operating environment of the handler and the environment in which the handler operates, i.e. the airport.

In this analytical model, it is assumed that m_1 is a function of schedule buffer time (${}_{G}S_{ij}^{b}$) and service efficiency of turnaround operations (denoted by m_2). Hence, the longer the schedule buffer time is (or the higher the operational efficiency is), the higher the efficiency of the schedule buffer time in absorbing

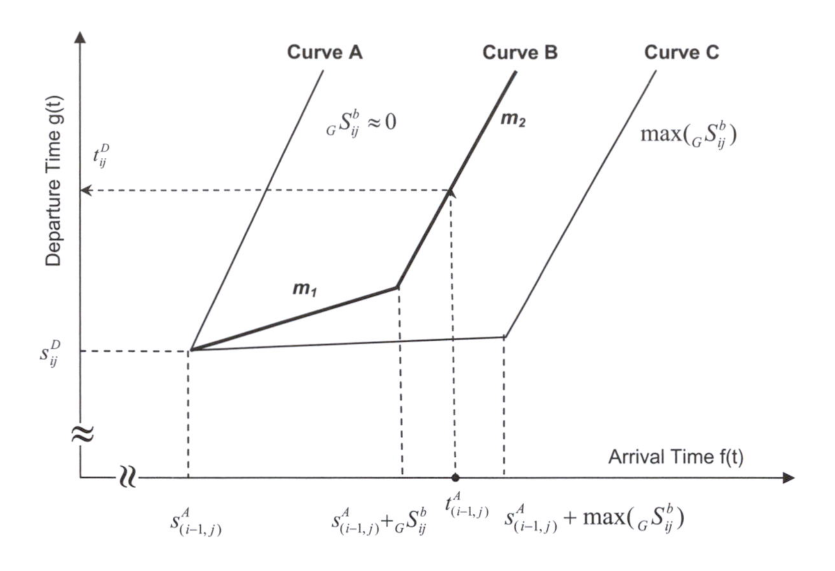

Figure 2.13 The relationship between arrival time *f(t)*, turnaround operational efficiency (m_2), and departure time *g(t)*

inbound delays, represented by m_1. Accordingly, a longer buffer time designed for a turnaround flight corresponds to a "flatter" curve due to smaller m_1 value such as the one illustrated by *Curve B*. Therefore, the relationship between inbound delay absorption efficiency (m_1), schedule buffer time (${}_GS^b_{ij}$) and ground service efficiency (m_2) is modelled as a piece-wise linear function by (2.2) below, corresponding to *Curve B* illustrated in Figure 2.13.

$$m_1 = m_2 \left[\left(\max({}_GS^b_{ij}) - {}_GS^b_{ij} \right) \Big/ \max({}_GS^b_{ij}) \right] \quad 0 \le {}_GS^b_{ij} \le \max({}_GS^b_{ij}) \tag{2.2}$$

where ${}_GS^b_{ij}$ is the schedule buffer time designed for flight f_{ij}; $\max({}_GS^b_{ij})$ is the maximum buffer time that absorbs inbound delays with approximately 100 per cent probability (or the maximum limit imposed by an airline); m_2 is a given parameter representing the operational efficiency of aircraft turnaround of a specific ground handler at a specific airport.

With the increase of schedule buffer time from zero to the maximum value, *Curve B* changes its shape from *Curve A* and approaches *Curve C* eventually when the schedule buffer time ${}_GS^b_{ij}$ is large enough. Since m_1 is also a function of m_2, when the operational efficiency of aircraft turnaround is improved, the parameter value of m_2 becomes smaller. Accordingly, the corresponding m_1 value decreases, and this results in a flatter *Curve B*.

With a defined mapping function, as shown in Figure 2.13, the departure time (t_{ij}^{D}) of f_{ij} can be formulated as a piece-wise linear function of the arrival time of $f_{(i-1,j)}$ ($t_{(i-1,j)}^{A}$) by (2.3 and 2.4) below, in which the operational characteristics of turnaround operations (modelled by m_1 and m_2) are described by (2.2) given above.

$$t_{ij}^{D} = m_1(t_{(i-1,j)}^{A} - S_{(i-1,j)}^{A}) + S_{ij}^{D} \qquad \text{if} \qquad S_{(i-1,j)}^{A} \leq t_{(i-1,j)}^{A} \leq \left(S_{(i-1,j)}^{A} + {}_{G}S_{ij}^{b}\right) \tag{2.3}$$

$$t_{ij}^{D} = m_1 * {}_{G}S_{ij}^{b} + m_2\left[t_{(i-1,j)}^{A} - \left(S_{(i-1,j)}^{A} + {}_{G}S_{ij}^{b}\right)\right] + S_{ij}^{D} \qquad \text{if} \qquad \left(S_{(i-1,j)}^{A} + {}_{G}S_{ij}^{b}\right) < t_{(i-1,j)}^{A} \tag{2.4}$$

where $t_{(i-1,j)}^{A}$ is the actual time of arrival of inbound flight $f_{(i-1,j)}$

$s_{(i-1,j)}^{A}$ is the scheduled time of arrival of $f_{(i-1,j)}$

s_{ij}^{D} is the scheduled time of departure of outbound flight f_{ij}

${}_{G}S_{ij}^{b}$ is the scheduled buffer time for turnaround of f_{ij}

t_{ij}^{D} is the actual time of departure of outbound flight f_{ij}

There is a one-on-one mapping between the arrival time variable ($t_{(i-1,j)}^{A}$) and the departure time variable (t_{ij}^{D}) because of the piece-wise linear transform function above. Since the actual arrival time is uncertain in real operations, the arrival time of $f_{(i-1,j)}$ is often modelled by a stochastic distribution, $f_{(i-1)}(t)$. Hence, the departure time distribution of f_{ij}, given the inbound arrival time distribution of $f_{(i-1,j)}$ by $f_{(i-1)}(t)$, is also a stochastic function, denoted by $g_i(t)$. Since the model given here by (2.3) and (2.4) is an analytical one (i.e. can be expressed by mathematical terms and operations), this model is also analytically solvable. In other words, we are able to obtain the exact function of $g_i(t)$, if we are given the function of $f_{(i-1)}(t)$ and the required parameters in (2.3) and (2.4).

2.5.4 Delay Cost Modelling – Passenger Delay Costs

Flight delays incur costs to affected passengers and airlines due to the time loss for passengers and the extra resources required during schedule disruption for airlines. Delays on the tarmac incur environmental costs to society because of extra fuel burn on the tarmac. To limit the scope of this model, the costs of delays only include delay costs to passengers and delay costs to airlines due to the loss of (aircraft) time.

Passengers who suffer delays lose time as well as the potential value of the delay time. In addition to the direct value of the loss of time, delays cause disruptions to passenger itineraries, business activities, and social arrangements, so the implications of delays to passengers have a profound social, business and personal impact. This has been the argument in Europe in recent years that has led to the amendment of previous legislation to further protect air passenger rights (European Union 2005).

In a report for the Performance Review Commission of Eurocontrol in Europe, delay costs to passengers were surveyed from both public sources and airlines (Eurocontrol 2004a). According to various sources and past studies of two European carriers, passenger delay costs were modelled and adjusted in detail. A simplified approach was to use the average hourly wage rate as a proxy to the losses a passenger may incur during delays (Wu and Caves 2000). Although, the "value of time" approach could be arguable in some cases, it provides a reasonably good proxy to passenger delay cost, when there are no better alternatives or quality data for modelling.

Passenger delay cost is modelled by (2.5) below. Passenger delay cost function, $C_P(d_{ij}^D)$ is modelled as a function of departure delay time (d_{ij}^D) and the function form of $C_P(d_{ij}^D)$ depends on the chosen marginal delay rate, i.e. $\gamma_P^m(d_{ij}^D)$. Although the delay cost function can be in any form in a more general expression (Tosic et al. 1995), for analytical simplicity and tractability, it is assumed in the TTA model that the marginal delay cost of passengers aboard is a constant, regardless of time. It means that the rate of delay cost ($\gamma_P^m(d_{ij}^D)$) does not change as the delay increases, and hence, is a constant in (2.5). As a result, the delay cost function $C_P(d_{ij}^D)$ has a linear form after integration.

$$C_P(d_{ij}^D) = \int_{d_{ij}^D} \gamma_P^m(d_{ij}^D)d(d_{ij}^D) \tag{2.5}$$

2.5.5 Delay Cost Modelling – Airline Delay Costs

When delays occur, airlines also incur extra costs due to aircraft operating expenses, crew costs, and expenses at airports, e.g. extra gate occupancy time. Various factors influence the cost of delays to airlines when delay occurs. To simplify the cost model we are building, only aircraft related delay costs are considered here, i.e. the "direct and hard" costs of delays. Accordingly, aircraft delay cost is defined as: *the hourly fixed operating cost per aircraft* (Wu and Caves 2000). Readers who are interested in other cost sources and relevant information can consult the report by Eurocontrol (2004) that provides sufficient details in this regard for the delay cost of an aircraft whether on the ground or airborne.

The direct cost of delaying an aircraft on the ground is modelled by (2.6), where $\varphi_{AC}^m(d_{ij}^D)$ is the marginal delay cost function of an aircraft. As previously in (2.5), aircraft delay cost $C_{AC}(d_{ij}^D)$ is modelled as a function of departure delay time (d_{ij}^D). The function form of $C_{AC}(d_{ij}^D)$ depends on the given marginal delay rate of an aircraft, i.e. $\varphi_{AC}^{m}(d_{ij}^{D})$. To simplify model building, it is assumed that the marginal delay cost of an aircraft is constant regardless of time. Although it can be argued that higher delays may incur higher costs for airlines due to aircraft usage, this model assumption is reasonable and consistent with our assumption regarding aircraft delay costs given earlier.

Based on this model assumption, the aircraft delay cost function $C_{AC}(d_{ij}^{D})$ has a linear form after integrating $\varphi_{AC}^{m}(d_{ij}^{D})$ in (2.6). If it is deemed necessary in the future to consider more delay cost factors, or to use a more complex form for $\varphi_{AC}^{m}(d_{ij}^{D})$, $C_{AC}(d_{ij}^{D})$ can be modified easily according to the model requirements.

$$C_{AC}(d_{ij}^{D}) = \int_{d_{ij}^{D}} \varphi_{AC}^{m}(d_{ij}^{D})d(d_{ij}^{D}) \tag{2.6}$$

2.5.6 Delay Cost Modelling – Airline Schedule Costs

To increase aircraft productivity, airlines tend to minimise the ground service time for turning around aircraft, so expensive aircraft time can be allocated to revenue-making block time. However, the use of the short turnaround time policy in an airline schedule is a double-edged sword. On the one hand, minimum aircraft turnaround time includes only minimum schedule buffer time which increases the risk of incurring flight delays when operational disruptions occur. Flight delays in a tightly connected airline network cause knock-on delays in a network, in which delays propagate via aircraft routing and resource connections (Wu 2005). The potential impact of knock-on delays is a multiplier effect that causes far more losses to airlines than the losses directly caused by the initial delays (Beatty 1998; Wu 2006).

On the other hand, a longer schedule buffer time included in the ground time reduces the risk of incurring flight delays, but compromises aircraft productivity and utilisation. Aircraft are expensive assets of airlines. Even for those airlines which do not financially "own" their aircraft, the leasing cost of an aircraft, e.g. B737, a popular type of aircraft for short-haul operations, was between US$100,000 and $150,000 dollars per calendar month in 2007.

There is a trade-off situation, where a shorter turnaround time (i.e. shorter ground buffer time) risks airlines having a higher probability of delayed flight departures but a high utilisation of aircraft; on the other hand, the use of a longer schedule buffer time reduces aircraft productivity, but maintains the desired operational reliability of airline schedules. Hence, the use of aircraft time as ground buffer time can be modelled as a form of "opportunity cost" in airline scheduling by $C_{AL}({}_{G}S_{ij}^{b})$ in (2.7) below, representing the cost for an airline to include schedule buffer time in aircraft turnaround operations in a flight schedule.

$$C_{AL}({}_{G}S_{ij}^{b}) = \int_{S} \delta_{AL}^{m}({}_{G}S_{ij}^{b})d({}_{G}S_{ij}^{b}) \tag{2.7}$$

As seen in (2.7) above, $\delta^m_{AL}({}_G S^b_{ij})$ is the marginal schedule time cost function which reflects the opportunity cost of flying flight f_{ij} by a specific type of aircraft. Often the larger the aircraft is, the higher the opportunity cost of aircraft time, reflecting the higher cost of owning and operating a large aircraft. It is realised from current practices in the industry that the schedule time opportunity cost increases in a non-linear fashion, increasing dramatically when the saved aircraft time is long enough for an aircraft to carry out another flight and earn additional revenues. This is often achieved in the re-assignment of aircraft routing optimisation, in which saved aircraft time of the whole fleet may change the optimal routing of aircraft, so that more flights can be accommodated in the network.

The expectation of extra revenues from conducting more flights is always the major financial incentive and driver for scheduling more flights in a network by reducing aircraft ground time, especially for LCCs. Accordingly, the marginal cost function $\delta^m_{AL}({}_G S^b_{ij})$ is assumed to take a linear form, meaning the marginal rate of schedule opportunity cost gets higher when schedule time usage is higher. Hence, the schedule time cost function, $C_{AL}({}_G S^b_{ij})$ in (2.7) is a function of ground buffer time ${}_G S^b_{ij}$ and takes a quadratic form after integrating the linear marginal cost function $\delta^m_{AL}({}_G S^b_{ij})$ in (2.7).

2.5.7 Turnaround Time Allocation (TTA) Model

A cost minimisation model, called the *Turnaround Time Allocation* (TTA) model, is developed as a tool to optimise the allocation of the turnaround buffer time $({}_G S^b_{ij})$ of a single flight (f_{ij}) in the context of the trade-off situation that: higher turnaround buffer time reduces the associated delay cost both for passengers and the airline, but higher schedule time increases the opportunity cost of using aircraft time in other revenue-making flight operations. The objective function of the minimisation model is given by (2.8). It seeks to minimise the total cost (C_T) in the model that includes the expected cost of delays (D_C) and the cost of schedule time (S_C). The trade-off condition is modelled by a weight factor, α which divides the costs and benefits between the two cost items.

$$C_T = \alpha D_C + (1-\alpha)S_C \qquad 0 \leq \alpha \leq 1 \tag{2.8}$$

The expected cost of delays, D_C as modelled in (2.8) above includes passenger delay cost, $C_P(d^D_{ij})$ modelled by (2.5) and aircraft delay cost, $C_{AC}(d^D_{ij})$ modelled by (2.6). Since both passenger delay cost and aircraft delay cost are functions of departure delay time (d^D_{ij}), both functions are stochastic cost functions with the same stochastic variable d^D_{ij}, representing departure delay. To account for the stochastic feature of delay and the associated costs, the cost of delays, D_C is modelled as the expected cost due to the departure delay.

The calculation of the expected cost of delay, D_C is formulated by (2.9), in which $g'_i(d^D_{ij})$ is the stochastic distribution of departure delay, d^D_{ij}. $g'_i(d^D_{ij})$ is a function transformed from $g_i(t)$, which is a function of the actual departure time of f_{ij}, i.e. t_{ij}^D as modelled below. Since $d_{ij}^D = t_{ij}^D - s_{ij}^D$ according to the definition of departure delay, the departure time PDF can be expressed by: $g_i'(d_{ij}^D) = g_i(t_{ij}^D - s_{ij}^D) = g_i(t) - s_{ij}^D$, because the scheduled time of departure, s_{ij}^D is a constant. S_C in (2.8) is a function of schedule buffer time, ${}_GS_{ij}^b$ and depends on the marginal schedule time cost function, $\delta_{AL}^m({}_GS_{ij}^b)$ as formulated by (2.7) earlier.

$$D_c = \int_0^\infty \left[C_P(d_{ij}^D) + C_{AC}(d_{ij}^D)\right] g_i'(d_{ij}^D) d(d_{ij}^D) \tag{2.9}$$

where $\quad g_i'(d_{ij}^D) = g_i(t_{ij}^D - s_{ij}^D) = g_i(t) - s_{ij}^D$

According to earlier assumptions and formulation, the TTA model is summarised as follows by (2.10). Equation (2.10.1) represents the weight factor used to model the trade-off between expected delay costs and schedule costs. (2.10.2) describes the opportunity cost of the schedule buffer time as a function of the decision variable, namely the usage of buffer time, ${}_GS_{ij}^b$. (2.10.3) is the cost of delays including passenger delay cost as in (2.10.4) and aircraft delay cost as in (2.10.5). The departure delay PDF is converted by (2.10.6) from the departure time PDF $g_i(t)$, which is a function (piece-wise linear transform function) of the inbound arrival time $t_{(i-1,j)}^A$, and the decision variable, ${}_GS_{ij}^b$.

Turnaround Time Allocation (TTA) optimisation Model:

To minimise: $C_T = \alpha D_C + (1-\alpha) S_C$ (with decision variable, ${}_GS_{ij}^b$) (2.10)

Subject to the following constraints:

$$0 \le \alpha \le 1 \tag{2.10.1}$$

$$S_C = C_{AL}({}_GS_{ij}^b), \text{ where } C_{AL}({}_GS_{ij}^b) = \int_S \delta_{AL}^m({}_GS_{ij}^b) d\, {}_GS_{ij}^b \tag{2.10.2}$$

$$D_c = \int_0^\infty \left[C_P(d_{ij}^D) + C_{AC}(d_{ij}^D)\right] g_i'(d_{ij}^D) d(d_{ij}^D) \tag{2.10.3}$$

$$C_P(d_{ij}^D) = \int_{d_{ij}^D} \gamma_P^m(d_{ij}^D) d(d_{ij}^D) \tag{2.10.4}$$

$$C_{AC}(d_{ij}^D) = \int_{d_{ij}^D} \varphi_{AC}^m(d_{ij}^D) d(d_{ij}^D), \text{ where } d_{ij}^D = t_{ij}^D - s_{ij}^D, \text{ and forms the departure delay PDF, } g_i'(d_{ij}^D) \tag{2.10.5}$$

$g_i'(d_{ij}^D) = g_i(t_{ij}^D - s_{ij}^D) = g_i(t) - s_{ij}^D$, where $g_i(t)$ is the departure time PDF with variable t_{ij}^D, derived from the following set of functions: (2.10.6)

$$\begin{cases} t_{ij}^D = m_1(t_{(i-1,j)}^A - S_{(i-1,j)}^A) + S_{ij}^D & \forall t_{(i-1,j)}^A, S_{(i-1,j)}^A \le t_{(i-1,j)}^A \le \left(S_{(i-1,j)}^A + {}_G S_{ij}^b\right) \\ t_{ij}^D = m_1 * {}_G S_{ij}^b + m_2\left[t_{(i-1,j)}^A - \left(S_{(i-1,j)}^A + {}_G S_{ij}^b\right)\right] + S_{ij}^D & \forall t_{(i-1,j)}^A, \left(S_{(i-1,j)}^A + {}_G S_{ij}^b\right) < t_{(i-1,j)}^A \end{cases}$$

where $t_{(i-1,j)}^A$ forms the arrival time PDF of inbound flight $f_{(i-1,j)}$, i.e. $f_{(i-1)}(t)$

$s_{(i-1,j)}^A$ is the scheduled time of arrival of $f_{(i-1,j)}$

s_{ij}^D is the scheduled time of departure of outbound flight f_{ij}

${}_G S_{ij}^b$ is the decision variable (scheduled buffer time of f_{ij})

t_{ij}^D forms the departure time PDF of outbound flight f_{ij}, i.e. $g_i(t)$

The inputs required for the TTA model in (2.10) include schedule data, the operational efficiency of aircraft turnaround at the study airport (from statistics), cost parameters (including passengers, aircraft and schedule time costs), and the arrival time PDF derived from historical data. The decision variable in optimisation is the length of schedule buffer time, i.e. ${}_G S_{ij}^b$ that is required to balance the trade-off between delay costs and schedule time costs. The weight factor in optimisation, α can be varied to reflect the consideration of weights in strategic airline schedule planning.

2.6 Schedule Optimisation and Case Study

The TTA optimisation model in the last section requires a number of input parameters before being able to solve the optimisation model and obtain results. Preparing model parameters is as critical and essential as building an analytical model. Hence, the following section provides guidelines and some details regarding the preparation of model parameters, through collecting relevant data, making appropriate assumptions, to calculating model parameters as input data. Results of optimisation based on numerical analyses are given for demonstration purposes and are compared with a case study at the end of this section that uses real airline data. To explore potential trade-offs between scheduling options and the impact on flight delays and operational robustness, a number of numerical analyses are conducted, following the numerical example of optimisation given.

It should be noted that although the approaches used in this section to derive some model parameters are generic in nature and are applicable in most situations, parameter values are calculated approximately from published financial data for the purpose of model demonstration only. Hence, parameter values are derived based on the needs of this model, with some simplification involved and subject to data availability. When the TTA optimisation model is adopted to conduct empirical studies, it is highly recommended that the user should review the parameter values and make appropriate adjustments accordingly. Model parameters can also be changed to study the impact of specific parameters on model outputs, i.e. sensitivity analysis.

2.6.1 Model Inputs – Passenger Delay Costs

When calculating the unit cost of delays to passengers, trip purposes and passenger characteristics are major factors that are necessary to explore. The literature on the value of time (VOT) of air passengers suggested that a passenger valued on-mode time at the wage rate for business flights and a quarter of the wage rate for leisure flights. Waiting and delay time were valued higher than on-mode time. In the report by Eurocontrol (2004), the costs of passenger delay to airlines were calculated based on two given delay cost figures provided by Austrian Airlines and an anonymous European carrier. The usage of VOT and the average hourly wage rate as a proxy to the unit passenger delay cost was discussed in the Eurocontrol report. Eventually, an (un-weighted) average of the two given cost figures from the two airlines was taken to represent the delay costs of passengers, being EUR 0.30 per passenger, per delay minute and per delay flight in Europe. A number of different projects provided different cost parameters based on different rationales and different research contexts (see Bates et al. 2001; DRI 2002; Eurocontrol Experimental Centre 2002; Institut du Transport Aérien 2000).

The differences in operating environments and the cost bases of serviced destinations in a network influence the calculation of the unit cost of passenger delay. Often the calculation of the unit cost of passenger delay itself is a challenging project, because the cost of delay involves both "hard" cost items, e.g. delay time and compensation, as well as "soft" cost items, e.g. loss of goodwill and damage of reputation. Hard cost items are easy to calculate, but soft cost items are hard to quantify. Without comprehensive studies on passenger delay cost, the most appropriate proxy to measure the cost of delay of passengers is the average hourly wage rate of passengers.

It is assumed earlier in the previous section that the marginal delay cost per passenger (denoted by $\gamma_P^m(d_{ij}^D)$) is constant, meaning the unit cost is the same for the first minute of delay and, say, the 30th minute of delay. Hence, γ_P^m is related to the average wage rate, flight classes chosen, trip characteristics and delay time perception of passengers. A survey by the Civil Aviation Authority (CAA) in the UK showed that the average wage rate was US $46 per hour for passengers using Heathrow Airport and US $42 per hour for passengers using Gatwick Airport (CAA 1996). On the other hand, business passengers using London City Airport exhibited a higher average wage rate of US $64 per working hour. The average wage rate for leisure passengers was US $39 per hour from the same survey by CAA in 1996. Since the context of the numerical study we are to conduct later is in European aviation, for the purpose of model demonstration and simplicity, the hourly delay cost of a passenger (γ_P^m in (2.10.4)) is assumed to take the average wage rate of US $42 per hour for the consideration of an average passenger during waiting/delay time at an airport (Wu and Caves 2000). It will be convenient to change the cost parameter in the future, if the study target passenger is in the business class or in a low-cost cabin/flight.

2.6.2 Model Inputs – Aircraft Delay Costs

Various parameter values of unit aircraft delay cost can be found in the relevant literature. Unit ground delay costs for European airlines presented in the literature were US $1330, $2007, and $3022 per hour for medium, large and heavy jets respectively (Janic 1997). The estimates of unit delay cost of an aircraft in the U.S. were US $430, $1300, and $2225 per hour with respect to small, medium and large aircraft in a study by Richetta and Odoni (1993). Although aircraft delay cost figures like these can be found easily in the literature, these parameters are often context sensitive, i.e. only valid within the context of the project and are not directly applicable to other cases without thoroughly reviewing the values. Hence, an empirical approach is needed to demonstrate how to derive an aircraft delay cost figure under the constraints of data availability and model needs faced by most practitioners in the industry.

When an aircraft is delayed at a gate either with the engines off or on, the airline not only incurs extra operating costs but also has to forego potential revenues. A comprehensive framework was given in the Eurocontrol report (2004) on the approaches to calculate aircraft delay costs and ownership costs. In many cases, the practitioners/modellers in airlines may not afford to conduct such a comprehensive project to obtain aircraft delay cost parameters, or some data required for calculation in the framework are not readily available. Accordingly, an empirical approach is introduced here to derive approximate cost parameters for both simplicity and convenience reasons.

The aircraft delay cost, denoted by $C_{AC}(d_{ij}^D)$ in (2.10.5), was previously defined as *the hourly fixed operating cost per aircraft hour*, which considered only the direct operating cost of an aircraft. Aircraft delay cost depends on aircraft types and sizes. For the purpose of demonstration, aircraft sizes are classified into three categories, namely medium, large, and heavy aircraft, as shown in Table 2.1. Aircraft operating costs of some selected airlines are calculated and listed in Table 2.2 by using published financial data from International Civil Aviation Organisation (ICAO) (ICAO 1997a; ICAO 1997b). It is noted that cost calculations in Table 2.2 were based on the average aircraft operating costs due to the unavailability of detailed cost breakdown with respect to aircraft types, fleets and sizes from published information. Practitioners and modellers in the industry usually have access to detailed cost breakdown by fleet types for their own airline. With the availability of cost breakdown details, it is highly recommended that the delay cost figure should be reviewed and calculated based on fleet types. The current empirical approach, however provides us with a proper methodology to derive an approximate value of the aircraft delay cost, when there are no better cost data available for calculation.

According to Table 2.2, aircraft operating costs were found to differ between carriers, and one of the reasons for this was due to the difference of fleet structure.

Table 2.1 Aircraft classification

Aircraft Classification*	**Maximum Take-Off Weight (MTOW, lb)**	**Average Seat Capacity**
Medium Aircraft (narrow-body jets)	MTOW $\leq$ 300,000	150
Large Aircraft (wide-body jets)	300,000$<$ MTOW $\leq$ 600,000	250
Heavy Aircraft (jumbo jets)	600,000$<$ MTOW	400

Note: * Classification based on Maximum Take-off Weight (MTOW) and average seat capacity.

Table 2.2 Hourly aircraft operating costs (with engines off at gates)

	British Airways (BA)	**British Midland (BD)†**	**KLM (KL)**	**Lufthansa (LH)**	**American Airlines (AA)**	**United Airlines (UA)**
Total Operating Expenses*	11,395	866	5,372	9,370	14,409	16,110
Aircraft fuel and oil expenses*	(1,150)	(50)	(580)	(1,014)	(1,726)	(1,898)
Subtotal+ Operating Expenses	10,245	816	4,792	8,356	12,683	14,212
Number of Aircraft	260	33	115	280	656	593
Aircraft Operating Costs ($/hr/AC)*	**4,498**	**2,822**	**4,757**	**3,407**	**2,207**	**2,736**

Notes: † British Midland is now known as bmi and bimBaby with an IATA code of WW. BD was the old IATA code when it was still British Midland. However, BD is used through out the numerical analyses to distinguish that parameter values are based on BD's data, and not WW's operation.

* Units in US $ (millions).

( ) represents negative values (deduction cost items).

+ Subtotal = (Total Operating Expenses)-(Fuel and Oil Expenses).

Sources: *Digest of Statistics, Financial Data Commercial Air Carriers, ICAO 1997* and *Digest of Statistics, Fleet-Personnel, ICAO 1997.*

For instance, British Airways operated 32 per cent heavy aircraft for long-haul intercontinental flights based on 1997 fleet structure (as shown in Figure 2.14) and consequently, had a high average operating cost of US $4,498. KLM operated proportionately more large jets than Lufthansa, so KLM had a higher average aircraft operating cost of US $4,757 than Lufthansa. Lufthansa had a similar aircraft fleet structure as United Airlines, but exhibited a higher operating cost of US $3,407 which was commonly seen as the cost difference between Europe and the U.S. American Airlines mainly operated large and medium aircraft and few heavy ones, so a lower operating cost of US $2,207 was reasonable. On the other hand, British Midland (which is now known as bmi/bmiBaby) used mainly narrow body jets and exhibited an hourly aircraft operating cost of US $2,822. Although cost breakdown information is commercially sensitive and often not available in the public domain, the study by Eurocontrol (2004, p. 58) provided a good guideline on the possible ranges of "aircraft block-hour direct operating costs" calculated from various industry sources.

2.6.3 Model Inputs – Schedule Time Costs

Airlines tend to minimise the turnaround time of aircraft in order to produce more revenue-making flight time and increase the utilisation of assets, e.g. aircraft and terminal/ground equipment (International Air Transport Association 1997; Eilstrup 2000). This is especially true for LCCs and those carriers which use intensive shuttle services or hubbing operations (Airports Council International 2000; Gittell 1995). Accordingly, an inference based on this reasoning is that aircraft ground time can be alternatively allocated to other flights as revenue-making airborne block time provided the minimum ground time for specific types

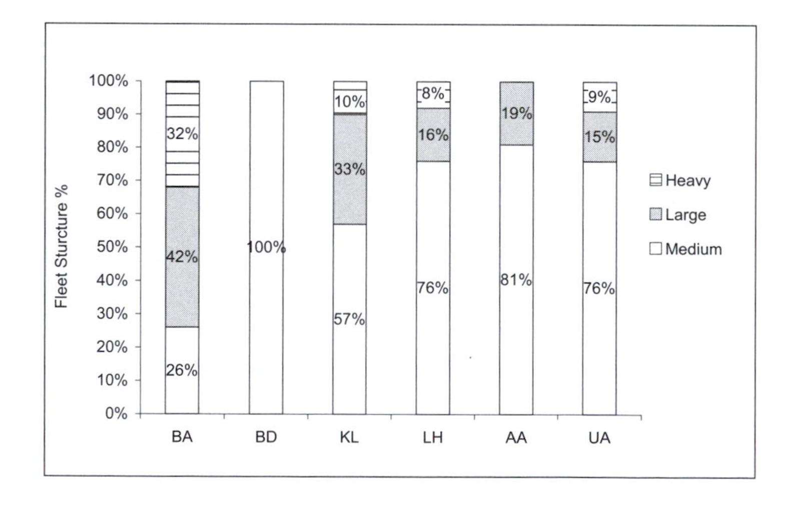

Figure 2.14 Aircraft types and fleet structure of selected airlines

of aircraft is met. In other words, the use of aircraft time as a buffer to control delays may improve the operational dependability of airline schedules and reduce expected departure delays, but the use of buffer time also incurs schedule time "opportunity costs". This trade-off is most obviously seen in LCCs, that reduce buffer time to the minimum allowable in order to increase aircraft utilisation and reduce schedule time costs.

To quantify the unit cost of airline schedule time (denoted by S_C in (2.10.2), some assumptions are required here. It is assumed that the variation of the fixed operational costs for an airline due to the variation of total block hours is insignificant, compared with the total annual revenues. In other words, it is assumed that the change of ground time for flights on the timetable causes only changes of revenues and variable operating costs, due to changes of available aircraft block hours. Based on this reasoning, the hourly schedule time opportunity cost is defined as *the marginal hourly operating profit of an airline* and is calculated by deducting hourly variable expenses from hourly revenues as demonstrated in Table 2.3 (based on earlier ICAO financial and fleet data).

Table 2.3 Hourly schedule time costs of selected airlines

	British Airways	British Midland	KLM	Lufthansa	American Airlines	United Airlines
Revenues*	12,226	890	5,699	9,986	15,856	17,335
Variable Costs* Fuel and oil	(1,149)	(50)	(580)	(1,014)	(1,726)	(1,898)
Maintenance	(663)	(64)	(350)	(441)	(937)	(1,049)
Station expenses	(1,602)	(93)	(875)	(1,434)	(2,102)	(2,195)
Passenger service expenses	(1,637)	(139)	(535)	(1,168)	(1,775)	(1,895)
Subtotal + (Revenues-Costs)	7,172	576	3,359	5,929	9,316	10,298
Flight Hours (hrs)	840,223	118,392	433,339	988,393	2,039,569	1,865,195
Schedule time costs ($/hr)	8,535	4,865	7,751	5,998	4,567	5,521

Notes: * Units in US $ (millions).
( ) represents negative values (deduction cost items).
+ Subtotal = (Revenues)-(Cost Items).

Sources: *Digest of Statistics, Financial Data Commercial Air Carriers, ICAO 1997* and *Digest of Statistics, Fleet-Personnel, ICAO 1997.*

It is observed from Table 2.3 that U.S. airlines had a lower average schedule time cost when compared with European carriers, except for the similarity exhibited between British Midland (bmi) and some U.S. carriers. Logically, the schedule time opportunity cost of heavy jets is higher than those of large and medium jets. This is supported by evidence illustrated in Figure 2.15, in which the schedule time cost of British Airways was higher than other airlines, because British Airways operated more long-haul flights with jumbo jets (denoted by the line of "holdings of large and heavy jets" in Figure 2.15), so a higher schedule time cost. Compared with British Airways, KLM operated more medium-distance flights, but KLM exhibited a higher schedule time cost than Lufthansa and two U.S. airlines, due to the usage of larger aircraft than other carriers. Figure 2.15 also suggests that the unit cost of schedule time can be further categorised according to aircraft sizes or even the stage length of flights, if more detailed financial information is available during the modelling processes.

2.6.4 Model Inputs – Arrival Time Distribution

One of the input parameters to the TTA optimisation model in (2.10) is the probability density function (PDF) for the arrival time of the inbound flight. Although it is convenient to assume that the arrival time PDF is normally distributed, there are many potential drawbacks in naively jumping to this assumption without thoroughly investigating historical flight data. Hence, some real flight data in 1999 from an anonymous European airline were collected to study potential arrival time PDFs in various conditions.

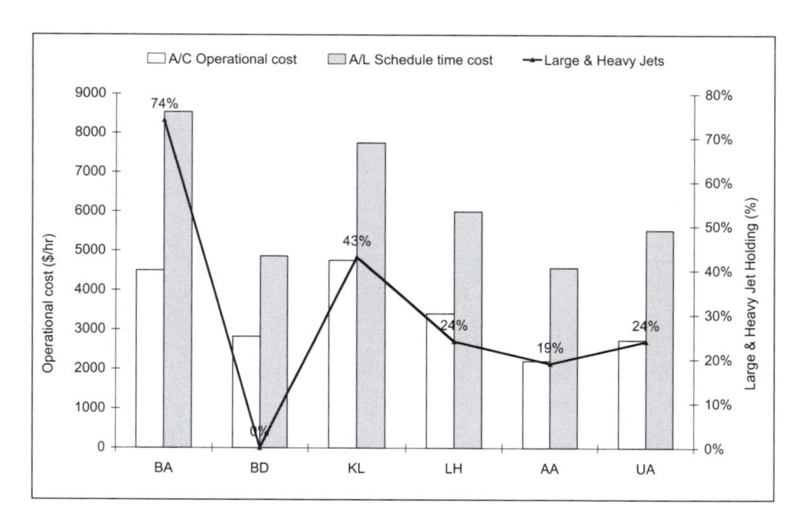

Figure 2.15 Comparison between schedule time costs and aircraft sizes

Various stochastic distributions were statistically tested to fit the real flight data and both *K-S test* and χ^2 *goodness-of-fit test* were used to ensure the "power" of curve fitting. Fitted probability curves are shown in Figure 2.16. Three different types of arrival patterns were identified and found to be representative for three different types of operations, based on a sample size of around 90 flights for each case (operated in the same season by the same carrier). Domestic flights within the base country showed a quasi-normal arrival pattern and were fitted by Beta (18,20) function. Short-haul international flights (within EU member countries), on the other hand, exhibited a right-tailed Beta (4,14) arrival pattern. Long-haul flights (inter-continental operations) showed a Beta (2,13) arrival pattern with a long right tail.

Based on the curve fitting results, the Beta function was chosen to model the PDFs of the arrival time of inbound flights, i.e. $f_{(i-1)}(t)$ in (2.10.6), because of its analytical form and tractability in calculations. Other types of stochastic functions can be used to model arrival patterns such as the Log-Normal PDFs used in a study from fitting historical flight data of some U.S. carriers (Lan et al. 2006). The drawback of using Log-Normal functions as the model input to the TTA model is that this function is harder to deal with analytically because of a complex function form, although this function is also analytically tractable. When a complex function like Normal or Log-Normal is chosen as an input to the TTA model, one will require to conduct numerical simulations using the Monte Carlo technique, in order to evaluate the optimisation model. On the other hand, the Beta function has a simple form and can be analytically calculated in the TTA optimisation model using a spreadsheet on computers.

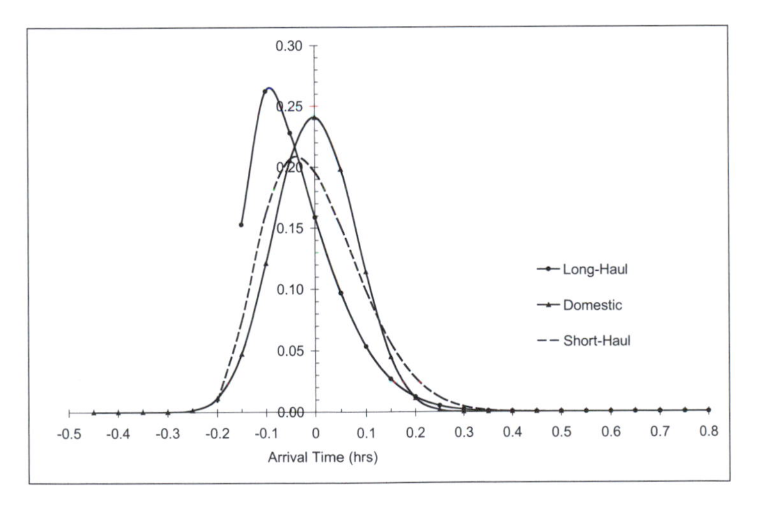

Figure 2.16 Arrival time patterns (PDFs) fitted from real flight data

Three types of arrival patterns, namely early arrivals, late arrivals, and normal-distributed arrivals, were employed in the following numerical analyses in this section, in order to study the influence of arrival punctuality of inbound aircraft on the departure punctuality of turnaround aircraft and the allocation of optimal schedule buffer time. For simplicity, Beta (10,3) was selected to model the late arrival pattern with the majority of flights (80 per cent) arriving within a maximum delay of twenty minutes (under a domain of 60 minutes on the x-axis) as shown in Figure 2.17. Only 20 per cent flights were punctual for the Beta (10,3) pattern. The Beta (3,10) distribution was used to model early arrival patterns, that had 90 per cent flights arriving within 14 minutes of delay and 30 per cent flights were punctual. Most arrivals modelled by Beta (3,10) had some delays, but these were relatively minor. The Beta (10,10) distribution was chosen to represent a quasi-normal arrival pattern, which had 55 per cent punctual flights with 90 per cent flights arriving within a short delay of 10 minutes.

The cumulative density functions (CDFs) of these Beta functions are illustrated in Figure 2.18. The "domain" of the chosen Beta functions is set to cover 60 minutes on the x-axis, meaning that the time range between the earliest and the latest possible arrival is 60 minutes. This is a simplification of real world cases, where an arrival delay can be as large as 120 minutes because of some very late flights. However, in analysing flight delays, it should be noted that causes of short delays are often quite different from causes of long delays (Eurocontrol 2004; Wu 2005). It is also a methodologically sound procedure to check these "sample outliers" in a statistical analysis, and for some cases, these outliers should be excluded from the main samples because of the possibility of biased results (for further suggestions on statistical analyses, readers can consult statistical texts for more information). To set the on-time arrival levels of Beta functions, we can shift the PDFs along the

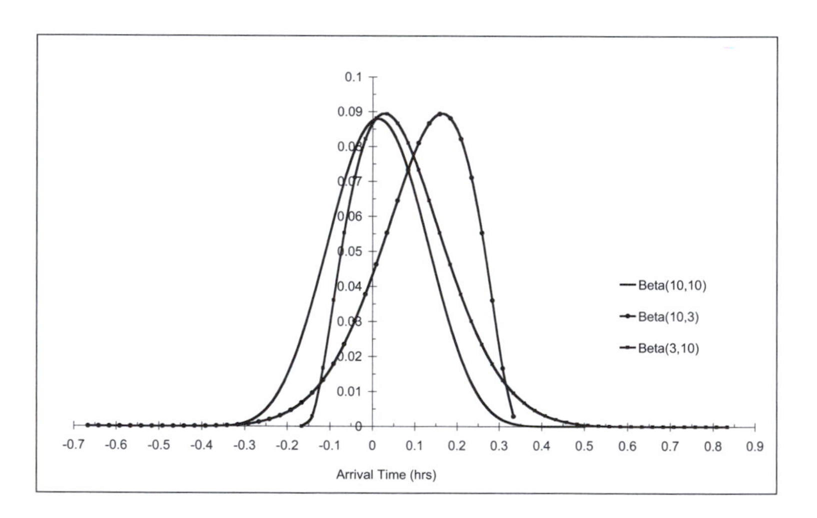

Figure 2.17 Various Beta functions (PDFs) used in numerical analyses

x-axis of Figure 2.17, so we are able to model arrival PDFs with various on-time performance levels and different delay patterns as demonstrated in Figure 2.18.

2.6.5 Optimisation Results

The parameter values used in the TTA optimisation model are summarised in Table 2.4 based on the empirical approaches given in earlier sections. Parameter values were derived from financial data published by the International Civil Aviation Organisation (ICAO), and were intended only for case study purposes and did not reflect the views of any studied carriers hereafter. To illustrate the application of the TTA model on different types of airline operations, British Airways (BA) and British Midland (BD) (now bmi) were chosen as study airlines, representing long-haul and short-haul operation focused airlines respectively. It should be noted that parameter values given in Table 2.4 are not meant to represent the real values of specific airlines and the choice of airlines in the following study was based on data availability considerations.

The Beta (3,10) function was used to represent the arrival time PDF of inbound flights for both cases. Ground operational efficiency, i.e. the m_2 parameter was given the value of 2, representing a scenario in which inbound delays can cause further disruption to ground operations. The TTA optimisation model was solved on a spreadsheet and results are given in Figure 2.19 and Figure 2.20 for BA and BD cases respectively. For the case of BA, we can see in Figure 2.19 that the expected delay costs (including passenger costs and aircraft costs) decreased when the schedule buffer time was increased. The linear form comes from the assumption made earlier, i.e. a constant marginal delay time cost, so a linear delay time cost function.

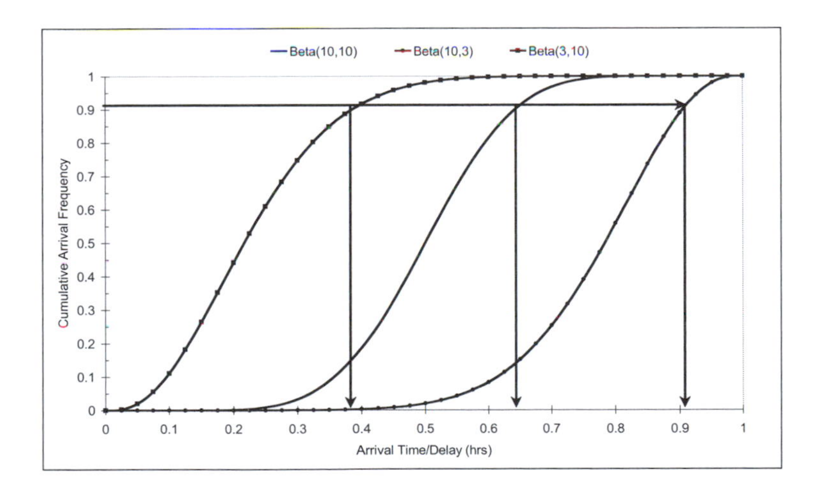

Figure 2.18 The CDFs of chosen Beta functions

Table 2.4 Parameter values used in numerical analyses

	Aircraft Delay Cost (\$/hr) ($C_{AC}(d_{ij}^{D})$)	Passenger Delay Cost (\$/hr) ($C_P(d_{ij}^{D})$)	Airline Schedule Time Costs (\$/hr) ($C_{AL}(_G S_{ij}^{b})$)	Ground Operation Efficiency (m_2)
British Airways[+] (BA)	4,500	5,880	8,535	2
British Midland[*] (BD)	2,822	4,100	4,865	2

Notes: [+] Average aircraft size for BA European flights is selected to be 200 seats with an average load factor of 0.7. The average hourly wage rate of British passengers is estimated to be \$42 per hour (Wu and Caves, 2000).
[*] Average aircraft size for BD is selected to be 150 seats with an average load factor of 0.65 (Wu and Caves, 2000).

On the other hand, the schedule time cost increased sharply as the use of buffer time increased. This was due to the assumption of a linear marginal schedule time cost in the model, so the schedule time cost followed a quadratic function as seen in Figure 2.19. Therefore, it can be observed from Figure 2.19 that the minimum of the total cost function occurred when the optimum schedule buffer time was set to be ten minutes. Hence, if the mean ground service time of a B767 was 45 minutes, then the optimal turnaround time for a B767 operation would be 55 minutes, which includes a 10-minute schedule buffer time.

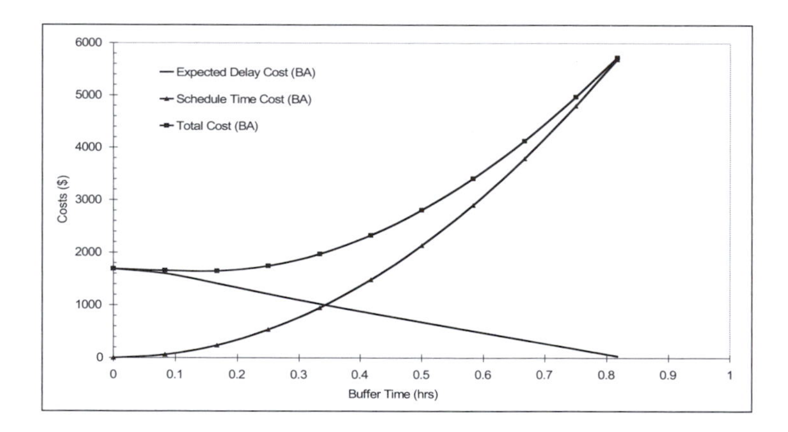

Figure 2.19 Cost curves of Beta (3,10) arrivals for the BA example

Compared with the BA example, cost curves in Figure 2.20 of the BD case displayed similar trends to the BA case, but the cost scale was significantly smaller. It can be seen from Figure 2.20 that the optimum schedule buffer time for this BD flight was ten minutes as well. Readers can also observe that the minimum of the total cost curve was influenced by both the “expected delay cost” and the “schedule time cost” curves. The total cost curve in Figure 2.20 was flat and concave, mainly due to the assumption of linear delay cost functions for both passengers and aircraft, compared with the quadratic schedule time cost function of airlines.

Under current model assumptions, it can be seen from Figure 2.20 that the schedule time cost increased more significantly than the improvement of expected delay cost with the use of schedule buffer time. The implication of this result validates the previous assumption that an airline is expected to save operational costs by optimising aircraft turnaround time. Also observed from Figure 2.20 is that the expected delay costs to passengers and aircraft were not as high as the cost from an airline’s schedule time opportunity cost. This observation explains why airlines focus on minimising aircraft ground time and improving the utilisation of fleets, because of the high cost of aircraft schedule time.

Current results here are based on model assumptions made earlier and are only valid under those estimated parameter values. For instance, the current results were based on a “low unit delay cost” scenario, in which the passenger delay cost took a linear form. If the delay cost function assumed a quadratic form (penalising long delays) or had a higher unit delay cost parameter (higher delay compensation for passengers), then the “expected delay cost” curve would have become concave, due to higher delay costs. This case would result in a higher delay cost than schedule time cost. Accordingly, under this scenario, airlines might be more willing to increase the schedule buffer in order to minimise operating costs. More

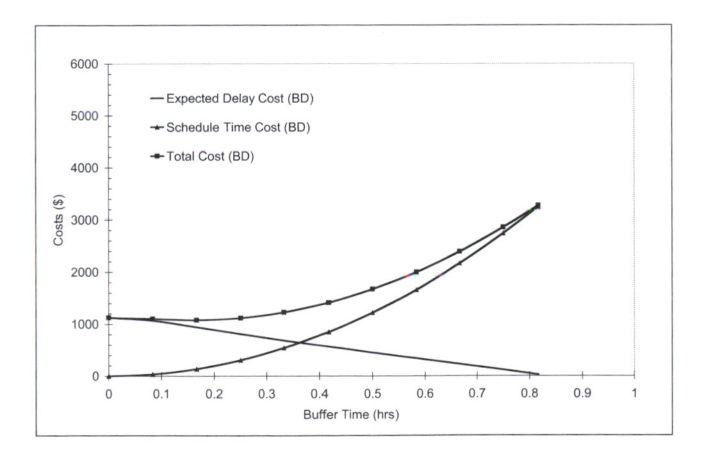

Figure 2.20 Cost curves of Beta (3,10) arrivals for the BD example

sensitivity analyses can be done to explore the potential influence of parameter values on optimisation results.

2.6.6 Case Study with Real Airline Data

A case study is provided here to demonstrate the application of the TTA model to a real case. Flight data collected from a European airline, denoted by Airline R, were used in the case study. Flight data represented three-month operations of two typical European city-pair flights, noted by RR-X and RR-Y, with turnaround operations at the base airport of Airline R. RR-X was scheduled to arrive at 18.45 and to depart at 19.45, representing a peak-hour operation. RR-Y was scheduled to arrive at 16.30 and to leave at 17.35 hours, which was operated during off-peak periods and had a long turnaround time. B757 aircraft were used to operate these two flights in the European operation of Airline R. To model the arrival patterns, Beta functions were fitted using available flight data. Both fitted PDFs passed the *K-S Goodness of Fit Test* and the Beta (4,9) and Beta (2,5) functions were chosen to model the arrival punctuality patterns of these two study flights. The on-time performance (OTP) (zero delays, i.e. D0) was 55 per cent punctual flights for RR-X and 60 per cent for RR-Y.

The TTA model was then employed to model the use of ground buffer time of flight RR-X and to explore the potential impacts this changes may make on the operational reliability of RR-X. The CDFs of the departure punctuality of RR-X from model results were shown in Figure 2.21. Different lengths of schedule buffer times were tested in the TTA model for RR-X and this resulted in various departure CDFs, corresponding to the various buffer times adopted. It is seen from Figure 2.21 that the longer the scheduled buffer time was in the turnaround time of RR-X, the more punctual RR-X departures would be. The observed departure

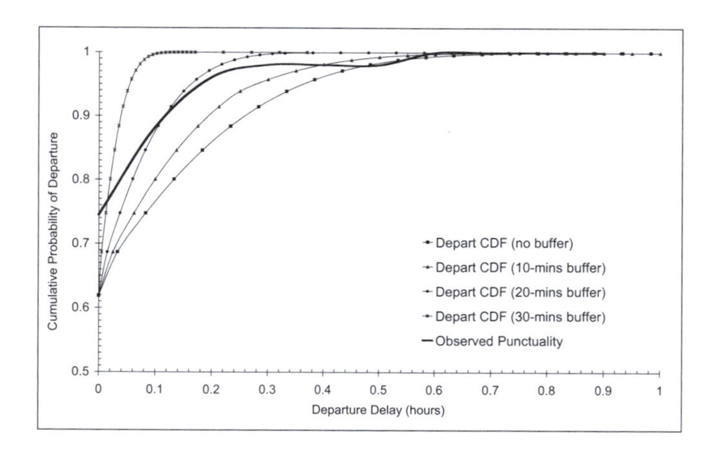

Figure 2.21 Departure punctuality of RR-X from observations and model outputs

CDF of RR-X was illustrated in Figure 2.21 by a thick solid line. It is seen in Figure 2.21 that the observed departure OTP of RR-X was close to the estimated departure CDF that had a schedule buffer time of 20 minutes.

The scheduled ground time of RR-X was 60 minutes and the standard turnaround time of a B757 aircraft by Airline R at the time of data collection was 40 minutes (for an European domestic operation). Consequently, the schedule buffer time was about 20 minutes, which was close to the model estimate here (also 20 minutes). However, it was also found in Figure 2.21 that the observed cumulative departure punctuality of RR-X was better with short departure delays (5 minutes) than model results and was worse than model results in some departures which had longer departure delays (more than 30 minutes). It was found from observations of the aircraft turnarounds of Airline R that long departure delays to turnaround aircraft resulted from both the long arrival delays of inbound aircraft as well as from operational delays due to disruptions to aircraft turnaround operations. As a consequence, a thicker right tail was found in the observed departure CDF of RR-X due to some extreme cases in observations that were not captured by the chosen inbound PDF, i.e. Beta functions.

To model extreme cases, one needs to adopt other types of stochastic distributions such as the Normal or Logistic distribution. However, other types of distributions may have a complex model form and this may cause some difficulties in the TTA model when trying to obtain an analytical solution. In such a case, one needs to resort to numerical methods such as statistical sampling and numerical simulation in order to obtain model solutions. The Monte Carlo method is a popular option and has been widely used in solving industrial problems (Fishman 1997).

The second case study was done by applying RR-Y's flight data to the turnaround model. The comparison between observed departure punctuality of RR-Y and estimated departure CDFs of RR-Y were shown in Figure 2.22. The observed departure CDF of RR-Y (represented by a thick sold line) developed closely to the estimated CDF that had no buffer time included. From the given flight schedule of RR-Y, it was known that the scheduled ground time of RR-Y was 65 minutes which included 25 minutes of buffer time when turning around a B757 aircraft. Model results showed that a 25-minute buffer time ought to be long enough to include 95 per cent of inbound delays to RR-Y. However, it was seen from Figure 2.22 that the turnaround punctuality of RR-Y was not commensurate with the amount of buffer time planned in RR-Y's schedule. In other words, the operational punctuality of RR-Y did not match the punctuality expected from RR-Y's schedule.

2.6.7 Strategies for Punctuality Management

In trying to understand this, a hypothesis we may investigate is that schedule OTP is endogenously influenced by airline scheduling philosophy, while exogenously affected by the operating environment in which an airline operates. In other words, the hypothesis states that it is feasible for an airline to manage its schedule punctuality by adjusting its flight schedule. As demonstrated in the case studies, RR-X exhibited good turnaround punctuality with respect to its scheduled turnaround time. On the other hand, the turnaround punctuality of RR-Y matched the estimated departure CDF which included no schedule buffer time, despite having a buffer of 25 minutes embedded in the schedule.

We can see from case studies that the turnaround time of RR-Y was not long enough to absorb potential delays from inbound aircraft as well as delays from aircraft ground operations. Yet, the endogenous schedule punctuality of a turnaround aircraft can still be achieved by good management of turnaround operations such as was observed with flight RR-X (note that the inbound OTP was similar for both RR-X and RR-Y). Hence, the observed schedule punctuality of RR-X fully reflects the amount of schedule buffer time included in its schedule.

It is usually argued by airlines that flight delays are mainly caused by uncontrollable factors such as air traffic flow management, passenger boarding delays, inclement weather and so forth. However, cases like flight RR-Y are not unusual for airlines and passengers. The case study of RR-Y offers airlines some clues to the better management of schedule punctuality. Managerial strategies to improve flight punctuality are therefore, recommended to focus on two aspects: flight scheduling, and the management of operational efficiency of the aircraft ground services.

It is feasible for an airline to manage schedule punctuality by optimally adjusting flight times in a schedule. For instance, flight RR-Y did not achieve its expected OTP, even though 25 minutes of buffer time had been scheduled in its turnaround time. Airline R, therefore can improve RR-Y's departure punctuality by scheduling longer turnaround time at the airport, if a longer ground time is needed, e.g. to accommodate connecting passengers, goods or crew. In addition,

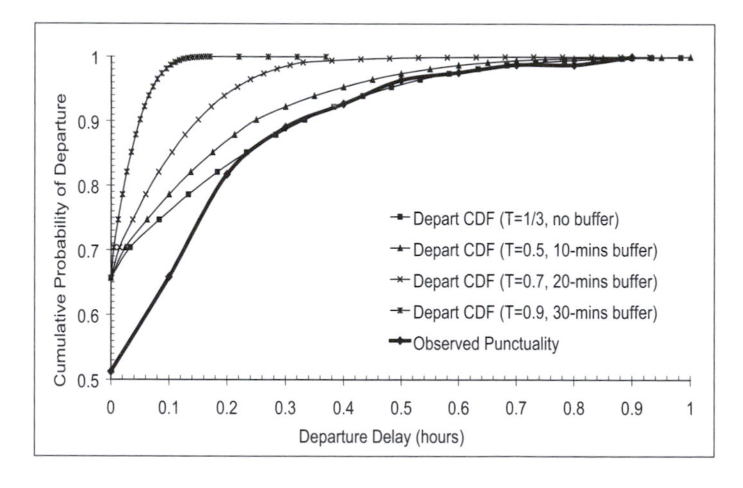

Figure 2.22 Departure punctuality of RR-Y from observations and model outputs

improvement of the arrival punctuality of the inbound flight to RR-Y can also help improve the departure punctuality of RR-Y, because less inbound delays often result in less ground operational disruptions and hence, potentially less departure delays for RR-Y. As a result, the departure punctuality of RR-Y can be improved by optimising flight times at the base airport of Airline R and outstations.

The management of schedule punctuality can also be achieved by improvements to the operational efficiency of aircraft turnarounds. It has been demonstrated earlier in this chapter how significantly the departure punctuality of a turnaround aircraft can be affected by the efficiency of aircraft ground services. Although short aircraft turnaround time increases the productivity of aircraft, it is also associated with a high likelihood of flight delays, affecting both airlines and passengers, because of the lack of delay absorption capacity in a tight flight schedule. On the other hand, the operation of aircraft ground services should be able to absorb some operational delays by operational means once it is realised that delays are likely to occur (Ashford et al. 1997; Braaksma and Shortreed 1971).

Most LCCs operate tight aircraft turnaround schedules at their base airports because the operational efficiency of aircraft turnarounds can be fully controlled and managed by these airlines. A longer turnaround time is often allowed at outstations, as less ground resources are available (especially at regional/secondary airports) and a longer ground time also allows some buffer against delays. Maintaining the efficiency of aircraft turnaround is believed to be the key factor for LCCs to deliver a reliable schedule of aircraft rotations (ACI 2000). However, there are still some potential risks for airlines operating tight aircraft turnaround. When schedule irregularities occur, the most effective solution to eliminate flight delays and delay propagation in an intensive aircraft rotation schedule is to cancel flights, which is often associated with high operating costs for airlines as well as inconvenience for passengers.

2.7 Summary

So far in this book we have treated airline ground operations at an airport as an aggregate process and based on this, we have developed a Turnaround Time Allocation (TTA) optimisation model, which allowed optimisation of the allocation of precious aircraft time in an airline schedule. An empirical model was also developed earlier that can be used by airlines to quickly adjust schedules, so to achieve the required OTP level in flight operations. Case studies and numerical analyses showed that inbound delays influenced departure delays significantly. In addition, ground operational efficiency, airline scheduling policy and the use of buffer time also influenced the operating OTP of flights. After examining two real flights, it was found that airlines could improve schedule OTP by two approaches: improving scheduling planning (optimisation) and managing the operational efficiency of airlines at airports, in particular aircraft turnaround operations.

Rather than studying airline ground operations on a "macro" level as we have in this chapter, in the coming Chapter 3 a micro perspective will be adopted to discuss the activities of airline ground operations as individual processes. Based on this view, a "micro" model will be developed. This model allows us to further explore the uncertainties involved in airline operations and how an airline may manage these uncertain factors in flight operations and scheduling, in order to achieve high operational reliability. Some widely used methodologies for modelling operational uncertainties by airlines will also be introduced in Chapter 3, including current practices on collecting delay data in the industry and some advanced models developed recently, aiming at improving our understanding of network complexity and mathematical modelling on airline operations.

Appendix
Notations and Symbols Introduced in Chapter 2

f_{ij}	flight *i* of route *j* that departs Airport A and arrives at Airport B (as in Figure 2.7)
$f_{(i-1,j)}$	the flight flown before f_{ij} on route *j* operated by the same aircraft
s_{ij}^{A}	the scheduled time of arrival of f_{ij}
t_{ij}^{A}	the actual time of arrival of f_{ij}
$s_{(i-1,j)}^{A}$	the scheduled time of arrival of $f_{(i-1,j)}$
$t_{(i-1,j)}^{A}$	the actual time of arrival of $f_{(i-1,j)}$
s_{ij}^{D}	the scheduled time of departure of f_{ij}
t_{ij}^{D}	the actual time of departure of f_{ij}
S_{ij}^{TR}	the scheduled turnaround time of f_{ij} at Airport A
S_{ij}^{BX}	the scheduled block time of f_{ij}
$d_{(i-1,j)}^{A}$	the arrival delay of $f_{(i-1,j)}$ and $d_{(i-1,j)}^{A} = t_{(i-1,j)}^{A} - s_{(i-1,j)}^{A}$
d_{ij}^{D}	the departure delay of f_{ij} and $d_{ij}^{D} = t_{ij}^{D} - s_{ij}^{D}$
d_{ij}^{A}	the arrival delay of f_{ij} and $d_{ij}^{A} = t_{ij}^{A} - s_{ij}^{A}$
d_{ij}^{OP}	the delays due to turnaround operations of f_{ij}
${}_{G}S_{ij}^{b}$	the scheduled ground buffer time of f_{ij} at Airport A
${}_{A}S_{ij}^{b}$	the scheduled airborne buffer time of f_{ij} en-route Airport A and B
$\hat{h}_{i}$	the realised (actual) turnaround time of f_{ij}

$f_i(h)$	the stochastic distribution of $\hat{h}_i$ with mean value, $\overline{h}_i$
$\hat{k}_i$	the realised (actual) flight time of f_{ij} between Airport A and B
$f_i(k)$	the stochastic distribution of $\hat{k}_i$ with mean value, $\overline{k}_i$
$f_i(t)$	the stochastic distribution of the actual arrival time of f_{ij}, i.e. t_{ij}^A
$g_i(t)$	the stochastic distribution of the actual departure time of f_{ij}, i.e. t_{ij}^D
$g_i'(d_{ij}^D)$	the function of departure delay; $g_i'(d_{ij}^D) = g_i(t_{ij}^D - s_{ij}^D) = g_i(t) - s_{ij}^D$
m_1	the efficiency of delay absorption by the scheduled buffer time of f_{ij}
m_2	the efficiency of turnaround operations at Airport A
$C_P(d_{ij}^D)$	the passenger delay cost function with a marginal delay cost function, $\gamma_P^m(d_{ij}^D)$; a function of delay time d_{ij}^D
$C_{AC}(d_{ij}^D)$	the aircraft delay cost function with a marginal delay cost function, $\varphi_{AC}^m(d_{ij}^D)$; a function of delay time d_{ij}^D
$C_{AL}({}_GS_{ij}^b)$	the opportunity cost of aircraft time with a marginal schedule time cost function, $\delta_{AL}^m({}_GS_{ij}^b)$; a function of ground schedule buffer time, ${}_GS_{ij}^b$
C_T	the total cost of the schedule time optimisation model, including the cost of delays, D_C and the cost of schedule time, S_C
D_C	the expected cost of delays including passenger delay cost, $C_P(d_{ij}^D)$ and aircraft delay cost, $C_{AC}(d_{ij}^D)$
S_C	the cost of schedule time calculated by $C_{AL}({}_GS_{ij}^b)$
α	the weight factor, representing the trade-off between delay cost and schedule time cost

Chapter 3
Managing Airline Ground Operations[1]

Chapter 3 focuses on the management of airline ground operations, especially on aircraft turnaround operations and passenger flow management at airports. First, the chapter starts by discussing some issues observed in daily airline operations and the complex resource connections between flights at airports. Second, the framework and techniques widely used by airlines to collect service data of aircraft turnarounds are introduced. The use of modern technologies to assist data collection and analysis is discussed with the introduction of the ACARS system used in the industry, and the development of the ATMS framework in this book.

In Section 3.3, some analytical models widely used by airlines and ground-handling agents for dealing with task monitoring and stochastic disruption management are discussed in detail. These models include a PERT model, which is powerful in task tracking and service planning, and a Semi-Markov Chain model, to be developed in Section 3.4, which better captures the stochastic factors in aircraft ground operations. Applications of mathematical models and the ATMS framework are provided throughout this chapter to demonstrate how these tools perform in real-world environment. Finally in Section 3.5, the strategies for managing passenger flows in an airport are explored from both the airport's and the airline's perspectives. Managerial and operational implications of efficiently and effectively managing passenger flows are examined, providing insight into airport retail revenues and airline passenger management in an airport terminal environment.

3.1 Issues in Aircraft Turnaround Operations

Aircraft turnaround operations refer to the activities conducted to prepare an inbound aircraft at an airport for a following outbound flight that is scheduled for the same aircraft. Accordingly, the activities of aircraft turnaround operation include both the inbound and outbound exchange of passengers, crew, catering services, cargo and baggage handling. Technical activities in turning around an aircraft include fuelling, a routine engineering check and cabin cleaning. Often turnaround operations for domestic flights are different from those performed for international flights due to differences in aircraft types, on-board service requirements and security requirements. Details of aircraft turnaround activities

1 This chapter is partially based on the following publications: Wu and Caves (2002); (2004); and Wu (2006); (2008).

have been discussed earlier in Chapter 2. Hence, this section focuses on the discussions of some potential issues surrounding aircraft turnaround operations. Further discussions on improving the operating efficiency of aircraft ground operations will follow in the next few sections.

3.1.1 Limited Turnaround Time and Schedule Constraints

Since passenger numbers and cargo/baggage loads vary from flight to flight and these numbers are only realised after the check-in is closed at the airport, the actual turnaround time of an aircraft is stochastic in nature. The *scheduled turnaround time* of an aircraft is defined as: *the time between the on-block and off-block time of the aircraft at a gate*. This scheduled turnaround time imposes constraints and operating pressure on airline ground operations, because delays to some service activities may cause delays to other services and eventually may result in departure delays. Transfer "traffic" may occur at airports during aircraft turnarounds such as transfers of flight/cabin crew, passengers and cargo/baggage between flights. The connection of airline resources (i.e. aircraft and crew), passengers and goods (i.e. baggage and cargo) can be significant for an airline that operates a hubbing network.

Under the complex resources connection mechanism among aircraft, passengers and crew, disruptions may occur to any of the processes of aircraft turnaround and may consequently cause delays to departure flights. Disruptions such as late connecting passengers, late connecting crew, missing check-in passengers, late inbound cargo/baggage or equipment breakdown are normally seen in daily airline operations. These disruptions occur randomly, although in real operations some flights may incur certain disruptions more frequently than others. The duration of the delays caused by disruptions is also stochastic. Accordingly, the impact of delays due to ground disruptions is uncertain, depending on the magnitude of disruptions and the nature of the network design. For instance, late inbound connecting passengers could be delayed by 15 minutes and cause brief delays for the outbound flight or flights in a multiple connection case. However, when connecting passengers are late by 45 minutes, this will cause significant delays to some outbound flights. The airline then needs to make a decision on whether or not to wait for the connecting passengers by holding departure flights or to leave without the connecting passengers.

While disruptions caused by air transport system capacity reduction attract much attention in the literature, mostly due to its large scale of impact (see: Arguello et al. 1998; Barnhart et al. 1998; Luo and Yu 1997; Rexing et al. 2000; Teodorovic and Stojkovic 1995; Yan and Young 1996), it is interesting to note that disruptions within this category account for roughly 40–50 per cent of total flight delays in Europe including those caused by weather (Eurocontrol 2004b; 2005). Other delay causes (the remaining 50–60 per cent) are contributed by airline operations and scheduling, in which reactionary delays may account for up to 20–30 per cent of the 50–60 per cent delay share; technical faults may account for up to 10 per cent (Eurocontrol 2004b; 2005). Delays cost money for airlines and passengers, regardless of the source of the delay. In this chapter our discussion will focus on those delay causes that airlines have better control during aircraft ground operations.

3.1.2 Resource Connections

Low-cost carriers are well aware how a fast and reliable turnaround operation can improve the bottom line of an airline business and the efficiency of an airline network through high utilisation of aircraft and low exposure to unexpected delays. Since delays may occur to any of the processes in schedule execution, buffer time is usually designed in flight schedules to accommodate unexpected disruptions and any consequent delays. Buffer time may be placed in the block time of flights as well as in the ground time for aircraft turnaround operations. Since airlines have more control and flexibility over the turnaround processes on the ground, the scheduled ground time for a turnaround is seen as a tactical and effective means to stabilise aircraft routing and to prevent further knock-on delays (also known as "reactionary delays" or "delay propagation" in the industry) through the rotation lines in an airline network.

Given the resource/passenger exchange among flights, disruptions to some resources/activities may cause delay propagation in the network via aircraft rotations, unless these delays are effectively contained by tactical measures by the airline such as flight cancellations, or absorbed naturally by the designed buffer time in the schedule. Hence, we can realise the crucial role played in maintaining efficient and effective aircraft turnaround operations in daily airline operations.

3.2 Collecting Service Data of Aircraft Turnarounds

3.2.1 Delay Data Collection

Flight data are collected by airlines in order to conduct analyses for operational improvements in the future. These data are also used to generate reports for relevant civil aviation authorities in different counties, to which airlines are required to report operational data, in particular delay statistics. Flight data are often referred to as the "OOOI data", which stands for "out of the gate, off the ground, on the ground, and into the gate". This is the standard procedure of flight operations from pushing back at the gate, taking off at the runway, landing at the runway of the destination airport and taxiing into the gate. The original purpose of this OOOI data was to trace the flight phases of an aircraft and maintain communication with an operating aircraft. In addition, the collected data may be used by some airlines for payroll purposes. A good history of flight data collection in the airline industry is available from Wikipedia (Wikipedia 2008a).

3.2.2 ACARS Data Recording System

The OOOI data is often collected automatically by avionic equipment aboard an aircraft, called the Aircraft Communication Addressing and Reporting System (ACARS). ACARS together with radio or satellite networks provide airlines and air traffic control authorities with a system to trace the flight phases of an aircraft during operations. Airlines compare the OOOI data with schedules and generate flight "delay data", which provides delay statistics as well as other flight-related information, e.g. carried passengers, and any operational disruptions to flights. In some countries, airlines are required to report these statistics (except those commercial data, e.g. passenger numbers) to civil aviation authorities. For instance, the Department of Transportation (DoT) in the U.S. requires airlines to report delay data, if they are operating within the United States and territories and have at least one per cent of total domestic scheduled-service passenger revenues, as described in 14 CFR Part 234 of DoT's regulation (DoT 2005). Airlines can also report voluntarily in the current regulation in the U.S. Monthly data are collected by various U.S. government agencies, providing information including flight delays, mishandled baggage, over-sales, consumer complaints and so forth.

In Australia, a similar regulation is in place that requires airlines to report the on-time performance for each route flown, subject to reporting criteria. The OTP statistics are reported for those routes where the passenger load averages over 8,000 passengers per month and where two or more airlines operate in competition. As of February 2008, there were 48 routes that met this definition in the Australian domestic network. Airlines also report overall monthly network performance data, representing over 99 per cent of scheduled domestic flight services in Australia (BITRE 2008). Similar regulations exist in the European Union and statistics are collected by the Central Office for Delay Analysis (CODA) of Eurocontrol (see Eurocontrol's web site for more details: http://web.hq.corp.eurocontrol.int/ecoda /portal/). Flight data (OOOI data) can be collected automatically by ACARS (or similar systems) or in a manual process by airlines. In a manual system, airline ground staff, pilots and ground handling agents will each record flight data separately, and these data are then centrally collated for delay analyses and reporting purposes.

3.2.3 IATA Delay Coding System

Together with the OOOI data, airlines also record causes of the delays. This information is helpful in determining the causes of delays and assists in improving airline operations and scheduling. The common framework for delay cause recording is the delay coding system developed by the International Air Transport Association (IATA) (IATA 2003). Delay causes are categorised into 12 major categories as shown in Table 3.1. There are 100 delay codes available for use, including some spare ones which can be adopted according to the individual airline's needs for data recording. Apart from the "numerical" coding system, each

Table 3.1 IATA delay codes summary

Delay Codes	Delay Categories
00–05	Airline Internal Codes
09	Schedules • Scheduled ground time less than declared minimum ground time
11–18	Passenger and Baggage • from Late check-in to baggage processing delays
21–29	Cargo and Mail • from documentation to late acceptance (mail only)
31–39	Aircraft Ramp Handling • from aircraft documentation late to fuelling & technical issues
41–48	Technical and Aircraft Equipment • from aircraft defects to scheduled cabin configuration adjustment
51–57	Damage to Aircraft & EDP/Automated Equipment Failure • from damage during flight operations to late flight plans
61–69	Flight Operations and Crewing • from flight plans to captain request for security check
71–77	Weather • from departure station to ground handling impaired by adverse weather conditions
81–89	Airport and Government Authorities • from air traffic services to ATC/ground movement control
91–96	Reactionary • from load connection delay to operations control
97–99	Miscellaneous • mainly industrial action with or outside own airline

numerical delay code is associated with an alphabetical code as well. For instance, delay code 11, (late passenger check-in due to late passengers) corresponds to the alphabetical code, PD.

Based on the IATA delay-coding framework, most airlines develop their own in-house delay coding systems to satisfy the demand of data collection. This demand often stems from the pressures of performance benchmarking and efficiency improvement among various units of an airline, e.g. the flight operations unit, engineering unit, or commercial unit. Therefore, some airlines have developed complex delay coding systems such as the one used by Air New Zealand (Lee and Moore 2003). An example of an in-house delay coding system of a carrier is given in Table 3.2. This system is based on the alphabetical delay codes and codes are categorised according to the IATA framework. However, unlike the IATA system, this airline reorganises the allocation of individual codes to meet its operational and managerial demands. In Table 3.3, the ZA category represents air traffic control (ATC) related delays. Within the ZA category, most codes are "8×" IATA codes, but a "93" code (aircraft rotation due to ATC) is also included. This reorganisation is more useful and meaningful from the airline's perspective, because it provides a better view of the true causes of delays according to key delay categories.

Table 3.2 An in-house delay coding system of a carrier

Codes	Description
ZA	ATC
ZC	CARGO
ZD	SECURITY
ZE	ENGINEERING
ZF	FLIGHT CREW
ZJ	INFORMATION MANAGEMENT
ZK	CABIN CREW
ZM	CATERING
ZO	OPERATIONS
ZP	CUSTOMER SERVICES
ZR	RAMP
ZT	TERMINAL OPS / DISPATCH
ZW	WEATHER
ZZ	AIRPORT + AUTHORITIES

Table 3.3 Sub-codes under the ZA code (continuing from Table 3.2)

Z Code	Delay Code	Code Number	Description
ZA – ATC	AC	81	Awaiting revised take-off slot
ZA – ATC	AE	83	ATFM restriction at destination airport
ZA – ATC	AM	89	Departure congestion inc. ATFM restriction
ZA – ATC	AMO	89	Multiple Push-back congestion, other operations etc.
ZA – ATC	AMT	89	Tow-on problem caused by GMC/ATC
ZA – ATC	AT	81	ATFM en-route demand/capacity
ZA – ATC	AW	84	ATFM weather at destination
ZA – ATC	AX	82	ATFM staff/equipment en-route
ZA – ATC	RAA	93	"RA" caused by ATC

3.2.4 Aircraft Turnaround Monitoring System (ATMS) Framework

Punctuality data are mostly compiled from the "time stamps" acquired manually by airline staff or automatically through ACARS or similar systems. Time stamps may include the take-off time (wheel off), landing time (wheel on), arrival time (on-block at gates) and departure time (off-block at gates), i.e. the OOOI data. However, operating data of ground handling, e.g. the catering unloading start time and finish time, are hardly as well recorded by airlines for the purpose of operations research. Airlines may have some records of these time stamps, but they are often scattered among different operating units and not collated centrally for analysis purposes.

For airline operations control, the widely available ACARS time stamps are used to track the flight phases of individual aircraft in daily operations. However, the lack of time stamps during aircraft ground time makes aircraft turnaround operations a "black box", and it is hard to co-ordinate available ground resources without relying on frequent radio conversations between operations controllers, airport duty managers and ground handling staff. The lack of operating data makes it difficult to evaluate the operating performance of ground handling services and also impossible to calibrate the operational procedures of different aircraft types at different airports.

Given the crucial role played by ground operations in controlling delays in an airline network, it would be of tremendous benefit for airline operations control if live operational data were available during aircraft ground operations on a real time basis. This would allow operation controllers to take precautions regarding potential events that might later on develop into delays for a departure flight, or potentially cause serious delay propagation in the network. The potential impact of this dynamic and real-time information on airline operations control has been well-demonstrated by the work by Abdelghany et al. (2004), in which flight delays and potential breaks of resource connections are projected ahead of schedule execution on a real-time basis.

Based on the needs of collecting data during aircraft turnaround operations, a turnaround monitoring framework is developed in this section, which serves as a platform to collect operational data, benchmark turnaround efficiency and calibrate the operational procedures of different aircraft types. Secondly, a real-time monitoring system is developed based on this framework, together with a data collection tool, providing all units involved in ground operations with situational awareness and up-to-date progress of aircraft under-going turnarounds. The Aircraft Turnaround Monitoring System (ATMS) is aimed at collecting operating data on a real-time basis during aircraft turnarounds. ATMS can also be used to conduct real-time monitoring tasks by utilising collected operating data. Based on the operational needs of individual ground handling units and the functional needs of operations monitoring, the ATMS framework is developed as an open framework as shown in Figure 3.1, which allows future development of add-on modules based on the same platform and structure.

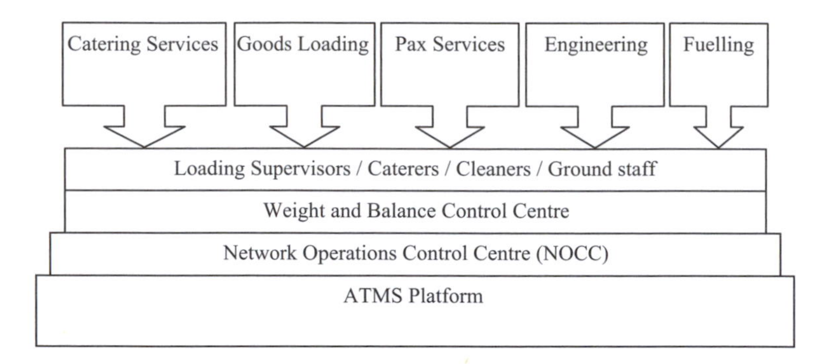

Figure 3.1 ATMS framework

Based on the main operational procedures of aircraft, turnaround activities are grouped into four major process flows, namely passenger, cargo, engineering check and catering in the ATMS framework. Activities within each process flow are chosen and included in the framework according to the needs of data collection and the importance of individual activities in turnaround operations. Using this framework, individual handling units only need to collect time stamps of key activities during turnarounds, so the progress of individual process flows can be easily shared between the handlers and the control centre, e.g. catering loading staff and the catering centre. Meanwhile, the collected information during turnaround operations can also be shared among different handling units, the ground operations centre at the airport and the network operation control centre at the carrier's remote headquarter. A list of key activities chosen in each process flow is given in Table 3.4. Some activities have an operational sequence to follow such as passenger, cargo and catering flows. Other activities such as refuelling and engineering checks are operated independently from other activities.

3.2.5 Implementation of the ATMS framework

The ATMS framework has been implemented using mobile devices with wireless telecommunication network technology, namely GPRS (a widely used mobile phone network service, which provides internet access to mobile phone users). Given the environment in which ground handlers work on the apron and to minimise the inconvenience of entering data, Personal Digital Assistants (PDAs) were chosen as the mobile device in this implementation. Activities and flows given earlier in Table 3.4 were programmed and implemented on a Palm PDA. Collected time stamps were both stored locally on the PDA and transmitted immediately on a real-time basis through the GPRS network to a remote database server.

The data flowchart of ATMS is shown in Figure 3.2. Multiple PDAs can be used for a single aircraft turnaround operation, if the ground handling strategy belongs to the "unit strategy", meaning different units independently conduct different jobs

Table 3.4 Activities modelled in the ATMS framework

Activity No.	Passenger	Cargo	Engineering	Catering
1	Position passenger steps/air bridge	Position cargo loader	Routine maintenance start	Open catering service door
2	Open passenger door	Open cargo door	Routine maintenance finish	Unload carts
3	Disembark passengers	Unload baggage	Fuelling start	Load carts
4	Onboard customs control/crewing	Unload cargo	Fuelling finish	Close door
5	Disembark crew	Load cargo	Wheel and tire check start	
6	Cabin and cockpit cleaning start	Load baggage	Wheel and tire check finish	
7	Final cleaning	Close cargo door		
8	Board crew	Remove cargo loader		
9	Crew check			
10	Board passengers			
11	Close passenger door			
12	Remove passenger steps/air bridge			

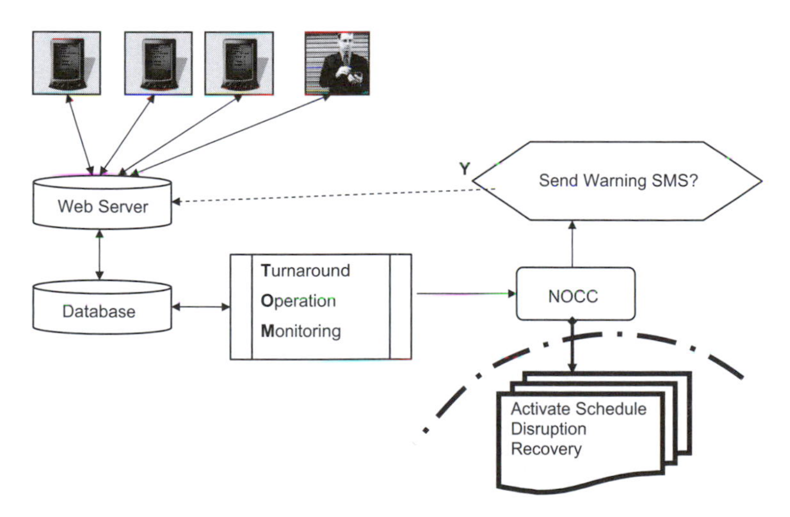

Figure 3.2 ATMS live data flowchart

in aircraft turnaround. If a "team strategy" is used for ground handling, then the team leader can use one PDA to control and monitor all the turnaround activities of an aircraft. An in-house real-time simulation model, namely Turnaround Operation Monitoring agent (TOM) was connected to the central database to monitor the status of multiple turnarounds at different airports (as long as there were live data input streams available), and updated in real time the estimated departure times of each monitored flight to each hand-held device on the apron.

Operation controllers at the Network Operations Control Centre (NOCC) of an airline may receive an automatic warning message from TOM, if the projected departure delay exceeds a predefined delay threshold. Operation controllers can then radio the manager of the ground handling unit to resolve the potential delay by operational means, or send text messages to the manager via PDAs and request proactive delay control actions. If the projected delay is long and cannot be resolved rapidly, the NOCC controllers have the option to activate the schedule disruption recovery protocol to deal with potential passenger itinerary disruptions, crewing disruptions and aircraft routing irregularities.

Two screen shots of the ATMS implementation on PDAs are given in Figures 3.3 and 3.4. The main menu of ATMS as shown in Figure 3.3 included six options: arrival, passenger, cargo, engineering checks, catering and departure. Arrival and departure options recorded the on/off block times of aircraft at gates, which were used as a reference to the ACARS arrival and departure time records. Activities under the "passenger" option were displayed on the PDA screen as illustrated in Figure 3.4. When an activity started/finished, the user only needed to click the activity on the screen. The current time stamp of the corresponding activity would be automatically obtained from the system time, stored and transmitted via the GPRS network to the remote database server immediately.

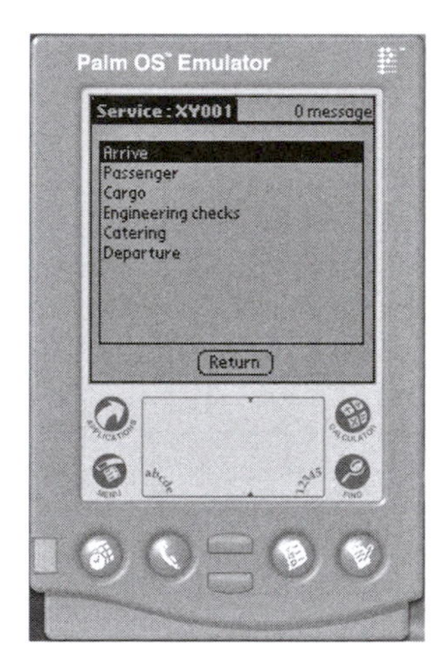

Figure 3.3 The main menu of ATMS of an example flight, XY001

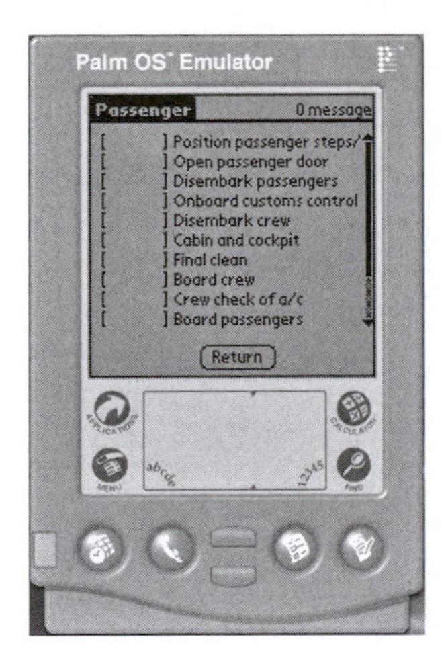

Figure 3.4 The input screen of the passenger processing flow of XY001

The ATMS system was first tested in April 2005 for remote communication functions. Immediately following the test was the trial at Sydney Airport to collect ground handling data to benchmark the operational efficiency of aircraft turnaround services. The second trial took place in January 2006 at Sydney Airport to collect data to improve the aircraft turnaround procedures for the B737 family aircraft. Due to the sensitivity of some data collected during the trials, only selected results are given here to demonstrate the ATMS framework and its implementation in real airline operations.

The catering service is given here as an example because it had caused some operational problems and departure delays in recent domestic operations of the Australian carrier. Figure 3.5 shows the observed start/finish times of catering services. The reference time in the following analysis is the "actual time of departure", i.e. the un-block time at gates. Those two bars highlighted with bold lines in Figure 3.5 represent the standard start/finish time of catering services for B737, which were 38 minutes and 25 minutes before pushing back the aircraft. We can see from Figure 3.5 that many catering services started early and this was mostly due to the long scheduled turnaround time and some early arrivals.

The scheduled turnaround time for this sample group ranged from 45 minutes to 65 minutes. 32 per cent of the catering services started late and these late services also caused 43 per cent catering services to finish late. The domino effect of this delay was to delay passenger boarding start time and consequently cause departure delays to 20 per cent of the total sampled flights, i.e. 16 out of 56 flights from our sample. However, catering service was not the sole reason causing departure delays to those 16 flights during our survey. Among them, six flights were also delayed due to passengers and goods connections between flights at Sydney Airport.

The goods unloading process usually causes little trouble unless the process is delayed due to late equipment allocation or equipment breakdown. The goods loading process, however, may cause delays to turnaround operations, in particular with complex goods connections among flights. Figure 3.6 shows the observed loading start time and finish time with respect to the actual departure time of flights. It shows 22 per cent of flights had late loading starts and this was reflected by the 17 per cent late loading finishes and consequent loading delays. Early loadings that appear in Figure 3.6 were due to long turnaround times. Overall, 21 per cent of sampled departure flights were eventually delayed due to loading related reasons. Among these delayed flights, half were delayed due to load connections specifically and the rest were due to late completion of goods loading.

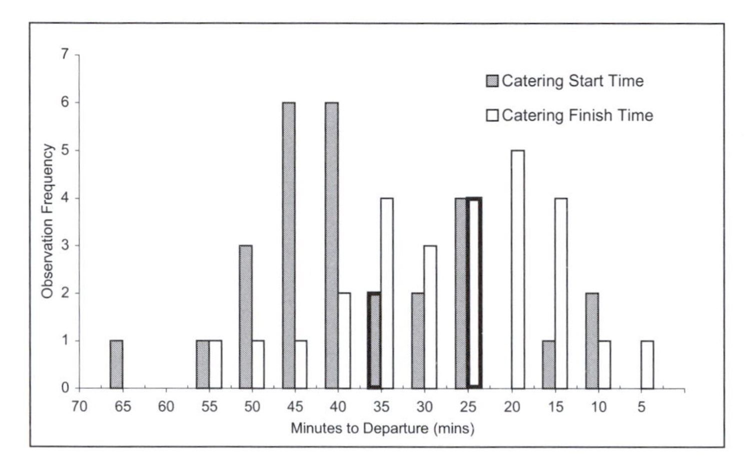

Figure 3.5 Start and finish times of catering service

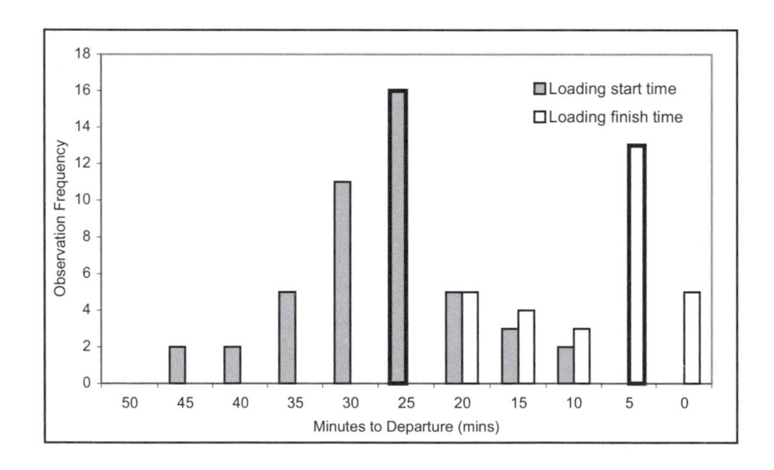

Figure 3.6 Start and finish times of goods loading

3.2.6 Implications for Airline Operations

3.2.6.1 Operations control, delay management and ground operation efficiency The availability of real-time turnaround operating data has significant implications for airline operations control and disruption management. The data captured by the ATMS system clearly shows the start and finish time of each turnaround activity. This real-time information improves the situational awareness of loading supervisors, airport duty managers and operations controllers. The further implication of the data availability is that operations controllers can take proactive actions to reduce departure delays and also mitigate potential delay propagation in the network.

The collected time stamps can be used to evaluate the operational efficiency of different procedures in turning around an aircraft, leading to improvements of the turnaround procedures, operational efficiency, and flight punctuality. Furthermore, the data gathered can be used as service quality indicators, which play an important role in establishing outsourcing contracts of ground handling with third party handlers. Without detailed operating data of turnaround operations, it would be hard to establish an objective service quality indicator to monitor the operational efficiency of ground handling, or the "service level" of ground handling. Lufthansa has moved towards this at its Munich and Frankfurt hub where Lufthansa does ground handling for other carriers (Mederer and Frank 2002; Schiewe 2005; Thon 2005).

3.2.6.2 Delay-coding systems Airlines use delay coding systems to record delay causes, so that in the future delays can be reduced by applying appropriate operational procedures. The standard IATA delay coding system consists of 100 delay codes representing different delay sources (IATA 2003). Apart from the IATA system, airlines also use in-house delay codes to encode further detail regarding the causes of delays, hoping that this information will help reduce future delays (e.g. Air New Zealand's delay coding system improvement by Lee and Moore (2003)). Given the complex involvement between the different groups involved in aircraft turnaround, a few operation co-ordinators and some radio conversations at a major airport are usually required to resolve delay code assignments for complex cases. The disadvantage of this practice, besides human resource costs, is the difficulty in determining the appropriate delay codes, which are truly "responsible" for delays. This difficulty also prohibits airlines from understanding the underlying causes of delays.

To solve this problem, some airlines use more than one delay code for flight delays. Although this seems to be a good method for tracing the causes of delays, this technique actually increases the complexity of delay code assignment and the following delay code analyses, because the delay-code trees may become too large to analyse. The developed ATMS system can significantly improve this process because operating time stamps make the delay root tracing an easy task for airlines. For instance, delays to passenger boarding finish time can be due to high passenger numbers (causing long boarding time), late start of boarding, late start/finish of cabin cleaning, late start/finish of catering service, late crewing procedures and late connecting passengers. Without clear information of individual turnaround activities, the delay code for this flight could have been assigned as "passenger boarding delay", which is in fact the consequence of delays instead of the root cause of the delays. This example demonstrates how collecting turnaround operating data can significantly improve airline operational efficiency as well as reduce delay propagation in an airline network.

3.2.6.3 Automatic data collection systems Although mobile computing devices can be used to collect operational data during aircraft turnaround operations based on the ATMS framework, one major drawback is that data collection still needs human interaction with mobile devices, adding extra burden to the ground handling agents, especially during peak operation hours and cold weather conditions (Thon 2005; Wu 2008). This manual data collection procedure can be semi-automated with the use of Radio Frequency Identification (RFID) tags. Until recently, RFID tags have mostly been used in the aviation industry for the tracking of baggage that has been listed on the agenda of "Simplifying the Business" by IATA (IATA 2008). Although the promotion and development of RFID use in airline and airport operations is not to replace the bar-coded boarding passes and bar-coded baggage tags in the short run, RFID has certain advantages for improving the operational efficiency of airlines and airports in areas such as the speedy processing of baggage sorting at hub airports and the high accuracy of tag reading.

Building on this RFID platform in airline operations, the data collection exercise of ATMS can be semi-automated. RFID chips can be attached to specific spots of the aircraft fuselage and RFID data readers can be mounted to ground handling equipment, together with wireless data transmission devices. For example, to collect service time stamps for cargo unloading and loading, RFID chips can be attached to a place near the cargo door, and RFID data readers can be mounted on the cargo loading trucks. When a cargo loading truck approaches the cargo door, the RFID reader automatically reads the information carried by the RFID tag near the cargo door; a time stamp of cargo unloading start time is obtained.

Often a data reader may continue reading tag information as long as a tag is within the rage of a reader. Hence, the last time stamp obtained becomes the finishing time of an activity. Based on this principle, the logging of most activities in aircraft turnaround operations can be fully automated including fuelling, baggage/cargo loading, and catering service. Some activities, however cannot be fully automated by RFID, e.g. cabin cleaning and passenger boarding. Time stamps for these services can then be manually collected on ATMS by ground staff, so this makes the data collection exercise semi-automatic.

3.3 Managing and Modelling Ground Service Activities

3.3.1 Project Evaluation and Review Technique (PERT) Model

To manage and model airline operations, the Project Evaluation and Review Technique (PERT) model is widely used in the industry. There are two goals of using PERT in managing airline operations: first, to evaluate and improve the efficiency of airline operational procedures; second, to improve the efficiency of airline ground resources allocation, especially human resources. In the 1970s, PERT and the other competing modelling technique, Critical Path Method (CPM) started being deployed in the aviation industry. Braaksma and Shortreed (1971) used CPM to model aircraft turnaround operations and the turnaround time an aircraft took occupying a gate in an airport terminal. The established model was able to closely describe current airline operations (relative to the time the study was performed) and even to predict and evaluate future changes to aircraft turnaround procedures and activities.

A later and more recent study by Hassounah and Steuart (1993) demonstrated how a similar concept, by considering stochastic flight delays and gate schedule buffer time, could lead to the improvement of overall performance of airport gate assignment. More recently, Adeleye and Chung (2006) developed a PERT model (in a network form) and used the CPM technique to calculate aircraft turnaround time with the assistance of the simulation software, Arena. The network-form of the PERT/CPM model was also used in a study of flight delay projection by Abdelghany et al. (2004) who used the "shortest route" algorithm to calculate the most likely path by which current flight delays may cause future flight delays due to delay propagation in an airline network.

Historically, PERT and CPM were developed independently in the late 1950s, albeit with a striking similarity between the two modelling techniques (Taha 1992; Wikipedia 2008b). Today the terms PERT and CPM often refer to the same technique, that is widely used in project planning, project scheduling, project controlling and management. In the following sections, "PERT" will be used to refer to the overall scheduling and management model, while "CPM" will be specifically used when we talk about the calculation of "critical paths" in a PERT model.

PERT is a modelling methodology used to describe the execution of a project that contains a collection of activities. PERT is also used to describe the "interdependencies" between some activities within a project. The interdependent relationships between activities usually take the form of a "chronological sequence" by which an activity cannot start until a preceding activity is finished. Mathematically, a PERT model is often represented by a "network diagram" which is composed of "nodes" and "arcs", representing individual activities of a project.

A "node" in the PERT network diagram represents an "event" which denotes a specific activity and the attributes of that activity, e.g. duration of the activity and variance of the duration. An "arc", on the other hand represents an interdependency between two nodes, starting from a "tail" node to a "head" node and is often represented by an "arrow" in a network diagram. Readers should note that we employ the "activity on node" (AON) convention in the following model, meaning that activities are represented by nodes and arcs are used only to represent interdependencies among nodes (Wikipedia 2008b).

A network diagram, Figure 3.7 is given below, representing some key service activities of aircraft turnaround operations. The list of the nodes and their corresponding turnaround service activities is given in Table 3.5. Node 1 represents the arrival of an aircraft at a gate, i.e. the "start node" and node 4 represents the status that the aircraft is ready for departure and pushing back from the gate, i.e. the "finish node". The path (1, 2, 3, 4) represents the workflow of cargo/baggage offloading and loading. The path (1, 5, 6, 7, 8, 9, 4) models the workflow of passenger disembarkation, cabin cleaning and passenger boarding.

Path (1, 10, 4) describes aircraft refuelling and (1, 11, 4) describes the aircraft routine maintenance check during aircraft turnaround operations. Branching nodes 12 and 13 represent catering offloading and loading procedures, assuming that node 5 precedes node 12, and node 13 precedes node 8. In practice, this means that catering off-loading only starts when passenger disembarkation is completed and passenger boarding will not commence until catering loading is finished.[2] The network diagram shown above is not necessarily indicative of the standard operating procedures of a catering service. For some types of aircraft (mostly narrow-body jets), passenger disembarkation/boarding and catering processing can take place simultaneously, if there are no potential physical conflicts between two procedures.

It should be noted that the list of activities given in this example is not necessarily a complete list of all activities involved in turning around an aircraft. International operations or operations by large aircraft often involve more turnaround activities than domestic operations by smaller aircraft. In addition, low-cost carriers often offer fewer "free" services aboard, so there are also less activities required in turning around an aircraft operated by a low-cost carrier. This is why the aircraft turnaround time of low-cost operations is often shorter than that of network carriers. To simplify the given example in this section, only major activities are modelled in our example problem.

3.3.2 Identifying Critical Paths in a PERT Model

One objective of applying the PERT model in project planning and management is to identify the critical path/paths of a project. A "path" in a network diagram of PERT consists of a series of activities that must be executed sequentially. An activity is said to be "critical" when the occurrence of any delays to this activity

2 This sequence is not always true for all types of aircraft. Depending on the location of galleys in the aircraft and the procedures of airlines, passenger handling can be conducted simultaneously with catering services, especially for short-haul low-cost operations.

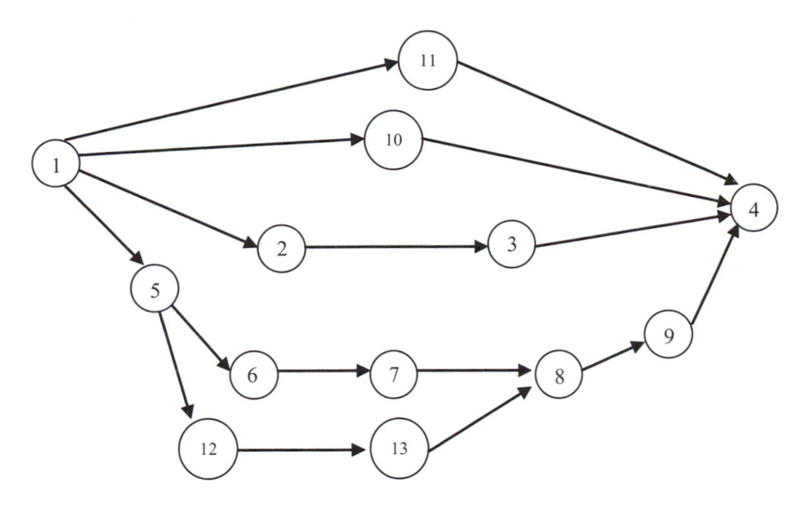

Figure 3.7 A network representation of key aircraft turnaround activities

Table 3.5 List of nodes and corresponding activities in a network diagram

Nodes	Activities	Duration (mins)
1	Arrival at the gate (Start)	0
2	Cargo/baggage off-loading	20
3	Cargo/baggage loading	25
4	Ready for departure and push back (Finish)	0
5	Disembark passengers and crews	10
6	Cabin cleaning	15
7	ATC flow control	3
8	Crew and passenger boarding	15
9	Flight operations and crew procedures	3
10	Aircraft fuelling	25
11	Routine maintenance check	20
12	Catering off-loading	10
13	Catering loading	10

results in the entire project being delayed. A path is said to be "critical" if the path consists of all critical activities in a project connecting from the start node to the finish node in the network diagram. This also means that any delay to activities on the critical path will cause delays to the entire project. Identifying the critical path has profound implications on project planning and management during operations. At planning stage, identifying the critical path can help resources allocation. During project execution, the knowledge of the critical path identifies those key activities that need the most attention and often more resources to ensure no delays occur to critical activities during actual operations.

The identification of the critical path includes two phases, namely the forward pass and the backward pass. The forward pass starts from the "start" node and is performed sequentially until the "finish" node is reached, which are node 1 and node 4 in our example. The forward pass calculates the "earliest start time" (denoted by ES_j) of activity *j* (at a node *j*), considering all connecting predecessor activities (denoted by *i*) and their corresponding "earliest completion time", denoted by EC_i. ES_j can be expressed by (3.1). The ES_j for the "start" node is zero and the EC_i is also zero for the start node. Following the forward pass, one can calculate the ES_j and EC_i of all nodes in a network diagram.

$$ES_j = \max_i \{EC_i\} \text{ where node } i \text{ precedes node } j$$
$$EC_j = ES_j + D_j \tag{3.1}$$

On the other hand, the backward pass starts calculating the "latest completion time" (denoted by LC_j) of activity *j* from the "finish" node to the "start" node. The LC_j time of the "finish" node is equal to the EC_j time on the longest path. The "latest start time" (LS_j) of activity *j* is expressed in (3.2), which is the LC_j time minus activity duration, D_j. The LC_i time of predecessor activity *i* is equal to the minimum LS_j time among all successor activities *j*. LC_i is expressed by (3.3). The "slack time" of an activity is the amount of time that the activity can be delayed without influencing or delaying the overall project duration. The slack time of activity *i* can be calculated by (3.4), once all other time indicators are available.

$$LS_j = LC_j - D_j \tag{3.2}$$

$$LC_i = \min_j \{LS_j\} \text{ where node } j \text{ are successors of node } i \tag{3.3}$$

$$Slack_i = LC_i - EC_i = LS_i - ES_i \tag{3.4}$$

The calculation of the earliest start time and the latest completion time provides information needed in determining whether an activity *i* is "critical" and lies on the critical path of a project. Activity *i* lies on the critical path if there is zero slack time for activity *i* and this path is the longest path in the network diagram. Readers should note that there could be more than one critical path in a complex project and the identification of critical paths is subject to the estimated activity times. In other words, when the service times of some activities are changed due to resource re-allocations or operational improvements during project execution, critical paths may change accordingly.

3.3.3 Managing Aircraft Turnaround Operations by PERT

The network diagram given earlier in Figure 3.7 was used to calculate the required indicator times according to the principles detail above. Calculated indicator times are shown in the network diagram in Figure 3.8, and is listed in Table 3.6. Each node (activity) has five time attributes including ES_i, EC_i, LS_i, LC_i, and $Slack_i$ which are organised in a designated format as shown in the legend key of Figure 3.8. We can see that the longest time required to finish the whole turnaround operation is 48 minutes. Those connected activities that had zero slack time were identified and formed a critical path, 1-5-12-13-8-9-4 (highlighted in the diagram).

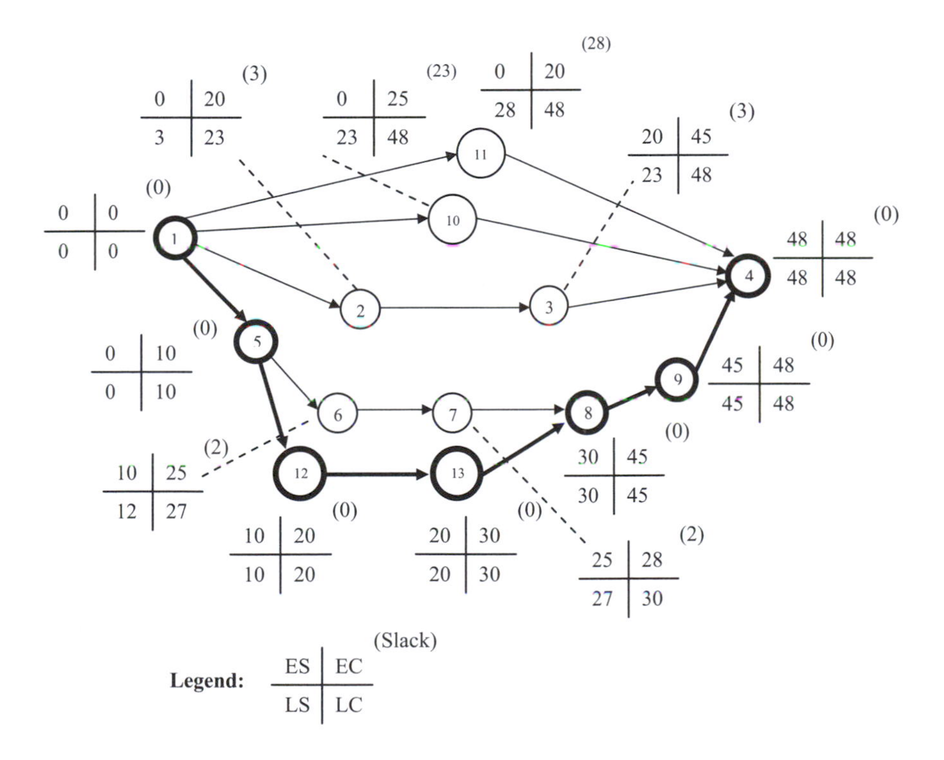

Figure 3.8 Activity times, slack times and critical path identification

Table 3.6 Time calculations for the example aircraft turnaround PERT model

Activity (node)	Duration (D_i, mins)	ES (mins)	EC (mins)	LS (mins)	LC (mins)	Slack (mins)
1	0	0	0	0	0	**0**
2	20	0	20	3	23	3
3	25	20	45	23	48	3
4	0	48	48	48	48	**0**
5	10	0	10	0	10	**0**
6	15	10	25	12	27	2
7	3	25	28	27	30	2
8	15	30	45	30	45	**0**
9	3	45	48	45	48	**0**
10	25	0	25	23	48	23
11	20	0	20	28	48	28
12	10	10	20	10	20	**0**
13	10	20	30	20	30	**0**

According to the calculation of slack time, one can find that the slack time for node 6 and 7 was only two minutes, meaning any delays longer than two minutes to cabin cleaning or due to air traffic flow control may potentially cause departure delays via a path other than the critical one. In addition, the slack time for cargo/baggage processing activities was only three minutes for off-loading and loading respectively. This means that the departure flight could be delayed, if delays to cargo/baggage processing exceeded the available slack time. On the other hand, there was more slack time available for aircraft refuelling (node 10) and routine maintenance checks (node 11), implying that these two services were generally not critical for the whole operation, unless excessive delays occur to these two services.

The presentation of a PERT model in the form of a network diagram is a clear way of presenting the interdependencies among various activities as well as viewing the complexity of the whole project. Often, a PERT model is given in the form of a *Gantt chart*, which is usually charted on a time scale for project management purposes. The network diagram given earlier was converted to a Gantt chart by using commercial software (OmniPlan on a Macintosh) and is illustrated in Figure 3.9. The assumed arrival time of the previous flight by the study aircraft was 8am. As one can see, the longest path (the critical path) took 48 minutes, with the aircraft being ready for pushing back at 08.48am and this path was also highlighted in the chart. On the Gantt chart, it was clear to see the

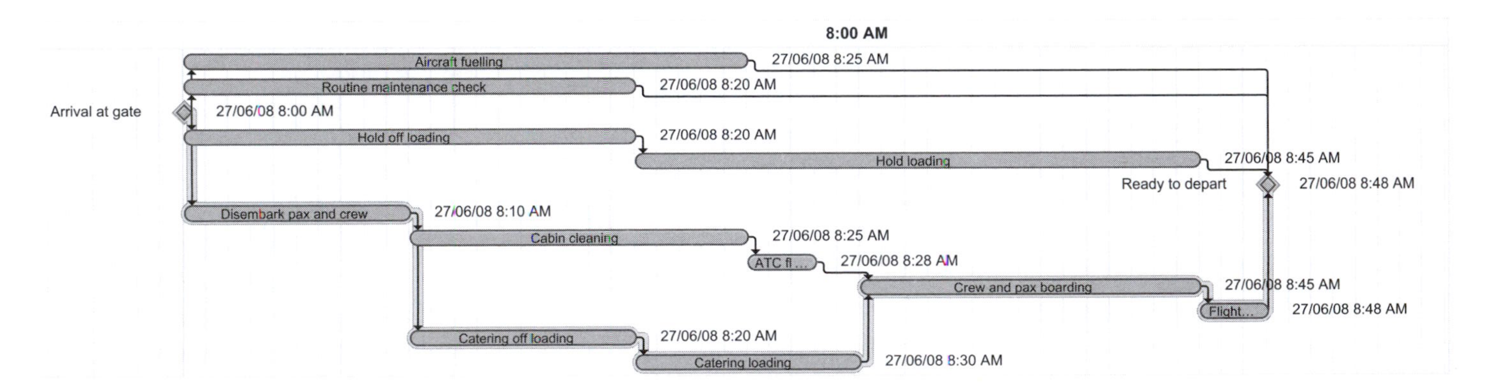

Figure 3.9 Gantt chart expression of the example turnaround PERT model

interdependencies among activities and the duration of an activity was reflected in the length of the bar in the Gantt chart, making the chart a convenient tool for real-time management of complex operations.

This Gantt chart can be converted into a network diagram drawn with the convention of "activities on nodes" as shown in Figure 3.10. This figure was drawn according to a similar convention as earlier in Figure 3.8, but now presenting only key information on nodes including activity name, earliest start time and earliest completion time. Although this new network diagram provides similar information to Figure 3.8, the inclusion of key information on a network diagram makes project management and critical path identification easier, especially if a project involves many activities with complex resources interdependencies among activities. These network diagrams can easily be created by commercial software.

3.3.4 Stochastic Activity Time in PERT

Very often in real world cases, the execution time of an activity is uncertain. Various stochastic forces may influence the execution time of an activity such as resources allocation and operational efficiency, making the completion time of a project stochastic as well. Given this modelling consideration, the uncertainties (or probability) of the service time of an activity are modelled by stochastic distributions such as the Beta distribution. To simplify time calculation in a PERT model that considers stochastic influences, the time estimate for each activity is based on three different time values:

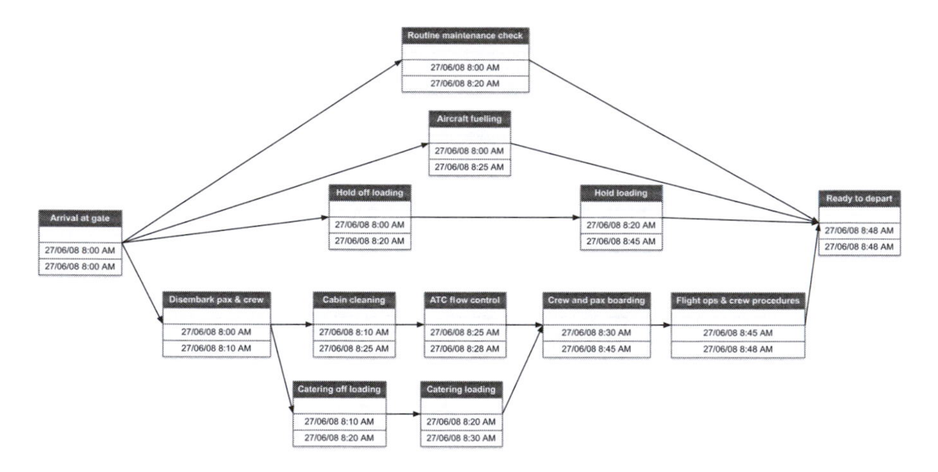

Figure 3.10 Network diagram of the example turnaround PERT model

a : denotes the *optimistic time*, representing the time when execution goes well
b : denotes the *pessimistic time*, representing the time when execution goes badly
m : denotes the *most likely time*, representing the time when execution is normally conducted

Using a Beta distribution, one can model the service time distribution of an activity with the unimodal point occurring at m with end points at a and b. The Beta function is not the only modelling choice in PERT, but the simplicity in time calculation and satisfactory approximation of the real situation makes Beta function a good candidate distribution for PERT models. Given the time estimates for individual activities, one can derive the mean value ($\mathrm{E}[D_i]$) and variance ($\mathrm{var}(D_i)$) of an activity based on the following equations (3.5 and 3.6) (Taha 1992).

$$\mathrm{E}[D_i] = \mu_i = \frac{a + b + 4m}{6} \tag{3.5}$$

$$\mathrm{var}(D_i) = \sigma^2 = \left(\frac{b - a}{6}\right)^2 \tag{3.6}$$

Given the use of stochastic distributions in PERT models, we can also estimate the probability of occurrence for each activity in the network. Let's assume that the occurrence of individual activities in a network is statistically independent, i.e. the time an activity takes is statistically *independent* from the other activities. For activity *i*, the earliest start time (ES_i) is the sum of a series of activities leading from the "start" node to node *i*. Since, each ES_i is a random variable (with mean, μ_i), according to the *Central Limit Theorem*, the sum of random variables is approximately normally distributed with the mean and variance as given in (3.7) and (3.8), in which *k* denotes the index of activities along the *longest path* leading from "start" node to node *i*. By strict mathematical modelling concepts, one needs to derive the exact distribution function of activity *i*, according to the different paths leading to a node. However, this is rather difficult, both in general and for real-world projects. Hence (3.7) and (3.8) below provide us with a reasonable estimation by applying the *Central Limit Theorem*.

$$\mu_i = \sum_k \mu_k \tag{3.7}$$

$$\sigma_i^2 = \sum_k \sigma_k^2 \tag{3.8}$$

Often it is required in project planning to calculate the probability of an activity occurring no later than a planned project milestone or deadline. For instance, we would be interested in estimating the probability that the ES_i time of activity i occurs no later than LC_i, the latest completion time of the activity. According to the *Central Limit Theorem*, this probability can be calculated by (3.9), where Z is the standard normal distribution with mean zero and variance one, and z_i is the test statistic of the milestone time. Readers who are interested in modelling details are urged to consult project management texts such as Taha (1992) and Bonini et al. (1997) for further details on PERT modelling.

$$P[\mu_i \leq LC_i] = P\left[Z \leq \left(\frac{LC_i - \mu_i}{\sigma_i}\right)\right] = P[Z \leq z_i] \tag{3.9}$$

To demonstrate how the uncertainties of activity times can be considered in a PERT model, we used the same example model from previous sections in the following demonstration. Uncertainties were considered by modelling the optimistic time, pessimistic time and the most likely time of an activity, i.e. the *a*, *b*, *m* factors in Table 3.7. The durations of activities (D_i) were kept the same as the previous example. According to (3.6), the variances of activities (Var_i) were calculated and listed in Table 3.7. These parameters were used to calculate the probability that a specific activity will occur no later than a specified milestone time according to (3.9).

Table 3.7 Parameter calculations of stochastic service times in PERT example

Activity (node)	Duration (D_i, mins)	*a*	*b*	*m*	D_i	Var_i
1	0	0	0	0	0	0.00
2	20	15	25	20	20	2.78
3	25	20	30	25	25	2.78
4	0	0	0	0	0	0.00
5	10	5	15	10	10	2.78
6	15	10	20	15	15	2.78
7	3	2	40	3	3	0.11
8	15	10	20	15	15	2.78
9	3	2	4	3	3	0.11
10	25	20	30	25	25	2.78
11	20	15	25	20	20	2.78
12	10	5	15	10	10	2.78
13	10	5	15	10	10	2.78

The corresponding probabilities of various activities in the PERT model were calculated against a milestone time, the latest completion time (LC_i). Specific activities were chosen to benchmark against other target times (shown by bold texts in Table 3.8) that had operational meanings. For instance, Activity 3 was evaluated against 48 minutes, the longest service time for the whole turnaround operation. Accordingly, the probability of Activity 3 showed that there is a 90 per cent chance that the cargo and baggage process will finish before it runs out of available slack time (3 minutes). Activities 7 and 13 were evaluated against their chosen target time, so we can see how likely it is that the start of Activity 8 will be delayed. Results showed that there is an 80 per cent chance that the cabin will be cleaned and ready for crew pre-boarding. Also catering preparation was likely to finish by the 35th minute with a 96 per cent chance. Accordingly, Activity 8 had an 82 per cent chance to finish within 48 minutes, before the cabin doors were closed and the pilots started requesting for aircraft push back. The overall processes of turning around this aircraft had only a 72 per cent chance of finishing within 50 minutes, if this was the planned aircraft turnaround time. Surely, if the planned turnaround time is increased, say to 55 minutes, then the probability of finishing turnaround operations would increase and be higher than 72 per cent.

Table 3.8 Completion probabilities against milestone times

Activity	**Longest Path**	**Mean**	**Variance**	**Milestone (LC_i)**	**Z_i**	**$Prob(Z<z_i)$**
2	(1–2)	20	2.78	23	1.80	96%
3	(1-2-3)	45	5.56	**48**	1.27	**90%**
5	(1-5)	10	2.78	10	0.00	50%
6	(1-5-6)	25	5.56	27	0.85	80%
7	(1-5-6-7)	28	5.67	**30**	0.84	**80%**
8	(1-5-12-13-8)	45	11.11	**48**	0.90	**82%**
9	(1-5-12-13-8-9)	48	11.22	48	0.00	50%
10	(1-10)	25	2.78	48	13.80	100%
11	(1-11)	20	2.78	48	16.80	100%
12	(1-5-12)	20	5.56	20	0.00	50%
13	(1-5-12-13)	30	8.33	**35**	1.73	**96%**
4	(1-5-12-13-8-9-4)	48	11.22	**50**	0.60	**72%**

3.3.5 Using Gantt Charts in Managing Aircraft Turnaround Operations

An empirical approach to managing aircraft turnaround operations in the airline industry is to apply standard operating procedures (SOPs) to turnaround operations. These SOPs are usually developed by the aircraft manufacturers and are modified by individual airlines to suit their local operational needs. SOPs are also different among different types of aircraft, depending on the aircraft engineering design and service needs. SOPs can be illustrated using Gantt charts and these charts are hence widely used in the airline industry. Gantt charts such as the one given in Figure 3.11 show a B737 turnaround SOP in Gantt chart format which is adopted by a carrier for short-haul turnaround operations.

The standard aircraft turnaround SOP benchmarks all tasks against the scheduled time of departure of a flight as shown in Figure 3.11. All tasks are required to be finished by the "latest finish times", so as to prevent resulting in knock-on delays to other tasks during aircraft turnaround or even delays to the flight departure. The interdependencies among tasks are not clearly shown on this Gantt chart, though. The principle used in the industry is that at any given time during the turnaround, all activities crossed by the time line on the chart can be operated simultaneously. For instance, at -25 minutes to departure time, a number of tasks are being executed simultaneously including: fuelling, unloading cargo/ bags, catering preparation, cabin cleaning and aircraft checks.

Since activities in aircraft turnaround have certain unique characteristics such as grouped processes and sequential workflows, the prevailing strategy to conduct ground handling in the airline industry is to assign individual "workflows" to different operating units, e.g. the catering unit, baggage/cargo handling unit

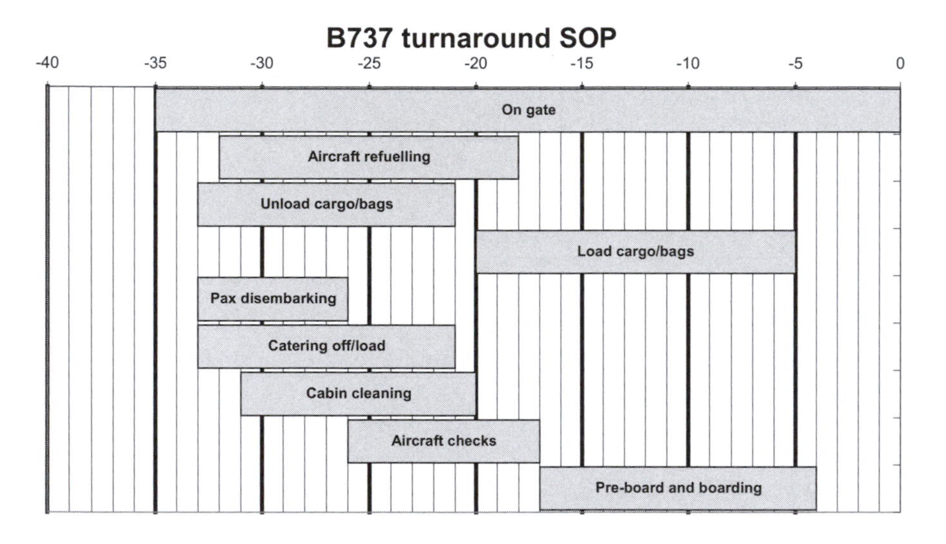

Figure 3.11 B737 SOP for short-haul turnaround operations

and passenger handling unit. This ground handling strategy, namely the "unit strategy", has the most benefits when the strategy is adopted by a home-base carrier or a third-party ground handling agent to handle the intensive needs of ground handling services, especially at hub airports where a significant portion of passengers are connecting passengers. However, the challenge of adopting this strategy is the need to ensure that good communication and co-ordination exists among operating units during operations.

On the other hand, some airlines, in particular low-cost carriers tend to use a "team strategy" to carry out aircraft turnaround operations. As suggested by the name, turnaround activities are conducted by a turnaround team, which is assigned to handle all turnaround activities of an aircraft (Gittel 1995; 2001). Accordingly, the handling team consists of multi-skilled staff and a team leader to ensure communication and co-ordination is well maintained during the operation. This team strategy requires a higher staffing level than the "unit strategy" and is more suitable for handling point-to-point traffic, i.e. handling basic loading and unloading duties, especially under time pressure. However, the gain from the higher operational efficiency and the capability to make up inbound delays may compensate the higher staffing cost or even the potentially higher delay costs due to delay propagation by tight aircraft rotation plans. Also, this team strategy is widely used in the car racing industry. A racing car such as a Formula One racing car would come to a pit stop during a race where a team of ground staff conducts a highly complex but well co-ordinated "turnaround operation" within a very short time, often less than ten seconds. Given the nature of highly co-ordinated turnaround tasks, it is not unusual to see errors in turning around a Formula One car in a race.

3.4 Managing the Stochasticity of Airline Ground Operations

3.4.1 Stochastic Disruptions and Modelling

The best way to understand the stochastic characteristics of airline operations is by studying real post-operation flight data. Table 3.9 below shows the statistics summary of six flights that were operated by the same aircraft during a one-day European operation of Airline P. Sample sizes of these six flights ranged from 47 flights to 146 fights, representing the operation of a season by the carrier. Due to data confidentiality, the identification of the airline and airports cannot be revealed in the following analysis. Table 3.9 reveals that the first flight of the rotation (flight #1) tended to incur delays due to loading problems (delay code #32) as well as early morning air traffic flow management restrictions in Europe (code #81). As the aircraft executed the rotation plan, delays from earlier flights tended to accumulate along the route, and aircraft rotation (code #93) appeared to be the most likely cause of delays according to statistics. The occurrence probability of code #93 for some flights was as high as 63 per cent (e.g. flight #5), which was nearly at the end of the route. The short turnaround time (20 minutes) scheduled for this flight was also a cause of its poor on-time departure performance. Air traffic flow management (code #81) caused frequent delays to this rotation plan as seen in the statistics in Table 3.9.

The summary statistics of Airline P's operation at two different airports, Airport A and B are shown in Figure 3.12. Statistics show that the operations of Airline P at Airport A had a higher chance of incurring delays due to passenger and baggage processing, aircraft handling at ramp, technical and flight operations, while it had a lower probability of incurring delays from airport/air traffic control and reactionary delays due to aircraft rotations. From these statistics, one can see the stochastic characteristics of airline operations, which are somewhat flight dependent, operating procedures dependent, and for some cases, also airport dependent.

If we compare the statistics of Airline P with Airline Q (another European carrier) operated at the same airport (Airport B), one can see from Figure 3.13 that the occurrence frequencies of certain delays were higher for one airline than the other. Flights of Airline P had a high probability of incurring reactionary delays due to aircraft rotation and flight schedules. On the other hand, Airline Q tended to have more issues with passenger and baggage processing and other operational disruptions from ramp handling, and airport/air traffic control. This figure shows again that the causes of stochasticity in airline operations and disruptions depend not only on the airline itself, but also on the operating environment (the airport and regional air traffic management), the nature of operations (more connecting traffic or more point-to-point traffic), and schedule planning (the usage of buffer time). While some stochastic disruptions are rather unpredictable in nature, e.g. weather, a good understanding of the stochastic forces involved in airline operations and borne with the operating environment will benefit not only airline operations and control, but also schedule planning and future improvement.

Table 3.9 Delay code frequency statistics

FLT	STD†	N	TR*	Delay Code A	Frq	%	Delay Code B	Frq	%	Delay Code C	Frq	%
1	530	143	–	32	16	11%	81	9	6%	18	9	6%
2	745	144	30	93	20	14%	81	17	12%	12	8	6%
3	920	144	40	81	61	42%	93	46	32%	89	2	1%
4	1045	146	30	93	85	58%	81	39	27%	87	3	2%
5	1305	146	20	93	92	63%	81	26	18%	32	9	6%
6	1555	47	60	81	20	43%	93	6	13%	85	1	2%

Notes: † Scheduled time of departure (STD) and * Scheduled turnaround Time (TR).

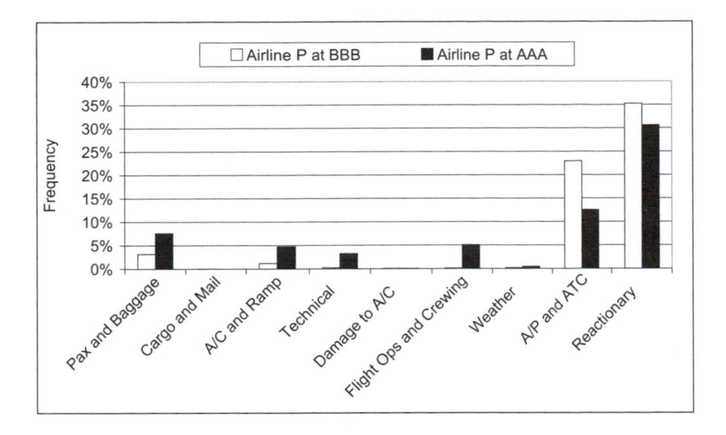

Figure 3.12 Delay frequency of airline P operation at Airport A and B

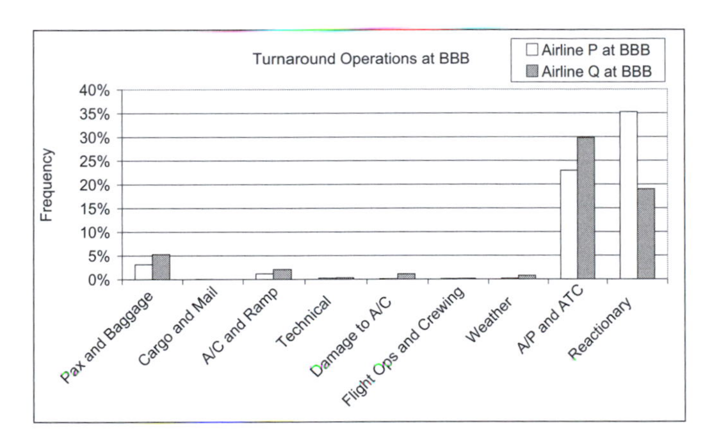

Figure 3.13 Delay frequency of airline P and Q operation at Airport B

The examples given previously are intra-Europe operations by two European carriers. Another example is provided here, which represents a carrier, denoted by Airline Z, that operates mostly inter-continental flights. A set of data was provided by Airline Z in 2005 which contained all flight data in the previous year, i.e. 2004. Delay causes were further grouped according to the in-house needs of Airline Z and the occurrence probability of delays due to each delay group was calculated and shown in Figure 3.14. As we can see, delays due to "reactionary" causes contributed 28 per cent to the overall operation in 2004, and there was a 10 per cent probability of incurring delays due to airport resources and airport authority issues. Passenger services caused only a 7 per cent chance of delays, while aircraft technical issues had a 6 per cent contribution to delays.

In addition, the mean delay time due to individual delay groups was calculated and noted in Figure 3.15. When the above figure was compared with Figure 3.14, a clear message was revealed. Although there was only a 6 per cent chance of incurring aircraft technical issues on the day of operation, the mean delay time of this group was as high as 132 minutes, nearly two hours. Some other frequent delay causes, e.g. reactionary and airport authority, only caused on average 73 and 18 minutes delay. In addition, delays due to airline operations (e.g. baggage and cargo processing) caused long delays, with an average of 37 minutes in 2004 operations.

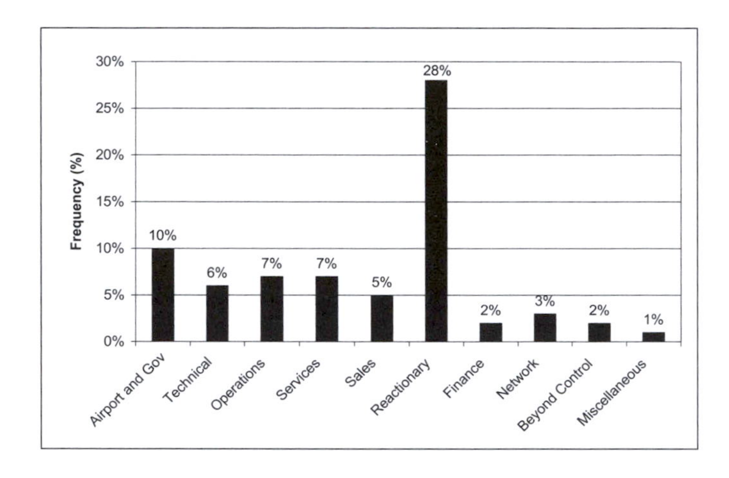

Figure 3.14 Occurrence probability of delay groups in 2004 by Airline Z

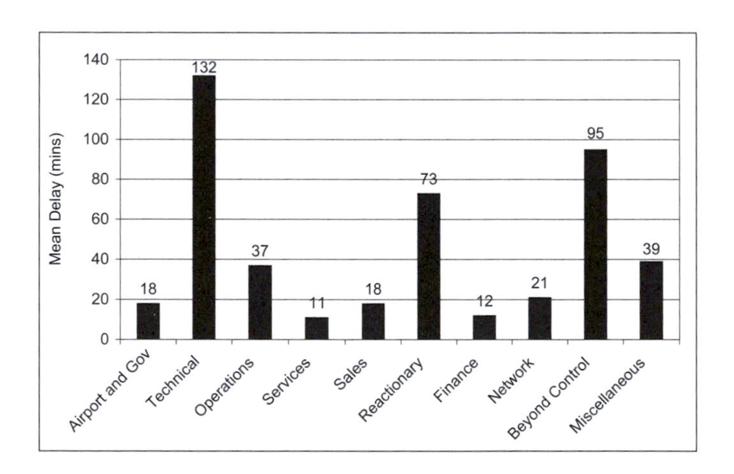

Figure 3.15 Mean delay times of delay groups of Airline Z

The lesson to learn from the example of Airline Z is that stochastic factors did significantly influence the operational robustness and schedule reliability in 2004. Technical problems, although they occurred rarely (6 per cent), caused high delays on average. This technical delay was equivalent in magnitude to the combined delays from all other airline operational causes, but the impact on the network was far greater. On the other hand, reactionary delays occurred quite often during operation in 2004 (28 per cent chance), and the average impact of reactionary delays was 73 minutes. Since most flights by Airline Z were inter-continental long-haul flights, ground operations involved more activities than domestic operations, and were subject to similar forces of reactionary delays in a network. From those examples provided in this section, one can appreciate how complex airline operations are, and the implicit connections between the forces of stochastic delays and schedule planning.

3.4.2 Semi-Markov Chain Model

On a micro-level, aircraft turnaround operations are comprised of a number of sequential processes as well as individual and independent activities. Each process and activity in a turnaround operation is subject to stochastic disruptions. Given the complex inter-relationships between processes in an aircraft turnaround operation, a micro simulation model is helpful and is able to achieve the following two objectives: first, the model should be able to describe the stochastic nature of individual service processes; second, the model should be able to capture the characteristics of aircraft turnaround operations, especially the sequential operating procedures such as unloading and loading goods.

Aircraft turnaround operations are comprised of a number of parallel workflows, which are conducted simultaneously at the airport ramp to turn around an aircraft for a following flight. Major workflows include passenger disembarkation and boarding (this also involves crew cabin check and cabin cleaning), cargo and baggage unloading/loading, catering unloading/loading and other independent work such as the aircraft engineering check and fuelling. A certain order for these procedures must be followed for those activities in the same workflow and delays to an activity may delay following activities in the workflow and possibly the departure time. For instance, cabin cleaning does not start until all passengers have left the aircraft and passengers will not start boarding until cabin cleaning is finished. However, in turnaround operations by some carriers, e.g. LCCs, it is often the case that flight attendants also carry out cabin cleaning (or at least some waste collection aboard) before landing. While passengers are disembarking via the front exit of an aircraft, cleaners may enter the aircraft via the rear exit and start cleaning up the cabin without waiting for the full disembarkation of passengers.

Given the sequential nature of activities in these workflows and the stochastic disruptions within workflows, the Semi-Markov Chains are used to model these workflows. Since the operating time of activities within workflows varies according to a few factors such as labour availability and work loading, the use of

Markov Chains is selected to reflect the stochastic aspects of aircraft turnaround operations. Disrupting events in workflows, e.g. missing check-in passengers or late baggage loading, are modelled as "disrupting states" in the Markov model with proper transition probability linking to normal operations, i.e. normal "states" in the Markov Chain. Since some activities in aircraft turnaround are conducted independently and not in a sequential order, event-driven simulation techniques are combined with the Markov model. More details of the semi-Markov model and implementation in real-world cases can be found in earlier publications of the author (Wu 2006; Wu and Caves 2004; 2002). Readers who would like to gain more information about Markov Chain and its recent applications can consult econometric or statistical books, see for example: Bose (2002) and Ching (2006).

3.4.3 Turnaround Simulation (TS) Model

Let t_{ij}^{D} be the actual time of departure of flight *i* on route *j* (denoted by f_{ij} and $\forall i \in F$), which forms a probability density function (PDF) of f_{ij} and is denoted by $g_i(t)$. s_{ij}^{D} denotes the scheduled time of departure of f_{ij}, so the departure delay of f_{ij} (denoted by d_{ij}^{D}) is defined by (3.10) as follows (this definition of delay is the same as the one used earlier in Chapter 2). New symbols introduced here are listed in the Appendix of this chapter.

$$d_{ij}^{D} = t_{ij}^{D} - s_{ij}^{D} \tag{3.10}$$

As formulated previously in Chapter 2, the actual time of departure of f_{ij} (t_{ij}^{D}) is a dependent variable influenced by two main factors, namely the actual time of arrival of the inbound flight $f_{(i-1,j)}$, denoted by $t_{(i-1,j)}^{A}$, and the stochastic turnaround operation time of f_{ij}, denoted by $\hat{h}_i$. $\hat{h}_i$ is the (longest) time required to finish all turnaround activities, including two major turnaround processes (passenger processing and cargo/baggage processing), delays due to disruptions, and other required aircraft service activities as formulated in (3.11) below.

$$t_{ij}^{D} = t_{(i-1,j)}^{A} + \hat{h}_i = t_{(i-1,j)}^{A} + \max\left[\hat{t}_i^{cargo}, \hat{t}_i^{pax}, \hat{t}_i^{events}\right] \tag{3.11}$$

The processes for the two major workflows of the aircraft turnaround operation are modelled by two Markov Chains, namely the cargo and baggage processing flow and the passenger processing and cabin cleaning flow. Service activities in each process are modelled as major "states" in the Markov Chain model according to the purposes of service activities such as goods loading and passenger boarding. The description of states in the workflow of cargo processing is given in Table 3.10. There is a main workflow for cargo and baggage processing, namely from State 1, State 2, State 3 to State 4, representing the stages from the arrival of an inbound flight, to the departure of an outbound flight (by the same aircraft). Operational disruptions to cargo and baggage processing are represented by State 5 to State 9 as illustrated in Figure 3.16 and described earlier in Table 3.10.

Potential disruptions to cargo and baggage processing include equipment failure, lack of labour, late check-in cargo, late check-in passengers, late baggage and so forth. The directions of arrows in Figure 3.16 represent the potential Markovian transitions between states.

Table 3.10 Cargo and baggage processing flow

States	State Descriptions	States	State Descriptions	IATA Delay Codes and Descriptions
1	Arrival			
2	Goods unloading	5	Cargo Processing	**22, 23, 26** Late positioning and preparation
		6	Aircraft Ramp Handling	**32, 33** Lack of loading staff, cabin load Lack of equipment, staff/operators
3	Goods loading	7	Cargo Processing	**22, 23, 26** Late positioning and preparation
		8	Aircraft Ramp Handling	**32, 33** Lack of loading staff, special load Lack of equipment, staff/ operators
		9	Passenger and Baggage	**11, 12, 18** Late check-in, check-in congestion Late baggage processing
4	Departure			

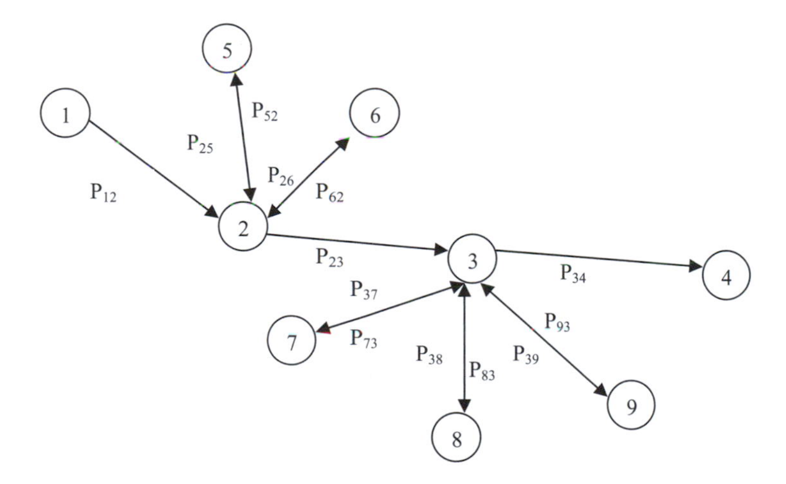

Figure 3.16 Cargo and baggage processing flow

$\hat{t}_i^{cargo}$ in (3.11) represents the time required to finish the cargo and baggage processing flow of flight f_{ij}. Activity k in the cargo workflow is modelled as a "Markovian state", so the cargo workflow is modelled as a Markov Chain in which each state may transit to some other states at time t with a state transition probability P_{pq}, i.e. the *stationary* probability of transition from state p to state q. This *Markovian renewal process* represents the change of states, including normal service activities and disrupting events that cause delays to turnaround operations. A total of K activities need to be carried out in this workflow and each activity k ($k \in \mathrm{K}$) has a stochastic operating time, namely the *state sojourn time*, denoted by $\hat{\phi}_{pq}^k$. The stochastic operating time of activity k (i.e. $\hat{\phi}_{pq}^k$) is modelled by a stochastic function, $\Phi_{pq}^k(\phi)$. The time spent in disruption "states" represents delays and these delays may or may not cause departure delay to f_{ij}, depending on the total time required to resolve disruptions and finish the turnaround operation. Hence, the actual time to finish cargo processing for flight f_{ij} is the sum of the stochastic service times of all K activities as in (3.12), assuming delays of different states are independent and additive.

$$\hat{t}_i^{cargo} = \sum_{k=1}^{\mathrm{K}} \hat{\phi}_{pq}^k \tag{3.12}$$

Similarly, the workflow of passenger processing is modelled as a Markov Chain in which the stochastic operating time of activity ω ($\omega \in \Omega$) is $\hat{\phi}_{pq}^{\omega}$ and the transition between states follows a Markovian renewal process with stationary transition probability, P_{pq}. The main workflow in this process is from State_1 to State _7, representing the stages from the arrival of an inbound flight, passenger disembarkation, cabin cleaning, passenger boarding to the departure of an outbound flight, as seen in Figure 3.17.

The state of air traffic flow management (ATFM) is included as State_4, because air traffic flow restrictions are often known to airlines in advance. In this circumstance, airlines will not start boarding passengers to an outbound flight until the ATFM delay is known and the projected departure time is near. Operational disruptions to passenger and cabin cleaning process are represented by State_8 to State_13 in Figure 3.17, which include missing check-in passengers, crewing problems, and flight operation delays during the departure procedure as detailed in Table 3.11. Hence, the actual time to finish passenger processing for flight f_{ij} is the sum of the stochastic service times of all Ω activities as in (3.13).

$$\hat{t}_i^{pax} = \sum_{\omega=1}^{\Omega} \hat{\phi}_{pq}^{\omega} \tag{3.13}$$

Some activities in aircraft turnaround operations are conducted independently from the two previous major workflows such as aircraft re-fueling and routine maintenance checks. These services may disrupt aircraft turnaround operations

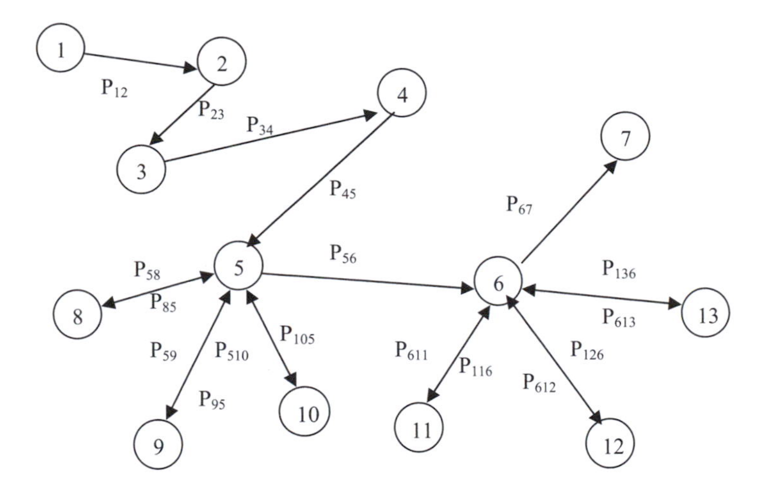

Figure 3.17 Passenger and cabin cleaning processing flow

Table 3.11 Passenger/crew/cabin cleaning process flow

States	State Descriptions	States	State Descriptions	IATA Delay Codes and Descriptions
1	Arrival			
2	Disembark Passengers and Crew			
3	Cabin Cleaning			
4	ATC Flow Control			
5	Crew and Passenger Boarding	8	Crew	**63, 94, 95** Late crew boarding, awaiting crew
		9	Passengers	**11, 12, 14** Late acceptance, late check-in
		10	Missing Passengers	**15** Missing check-in passengers
6	Flight Operations and Crew Procedures	11	Flight Operations	**61, 62** Flight plan, operational requirements
		12	Departure Process	**63, 89** Airport facilities, ground movement
		13	Weather	**71, 72** Weather restriction at O/D airports, Removal of snow/ice/sand
7	Departure			

when these services are delayed and the finish time of a service is later than the scheduled departure time of a flight. In addition, some disrupting events, e.g. aircraft damage during turnaround and un-scheduled aircraft change may cause delays to aircraft turnaround. Since these disrupting events and the consequent delays occur independently and unexpectedly, these events are modelled as independent stochastic events.

Four major disrupting events are modelled here, namely disruptions to refueling, engineering check delays, aircraft damage, and aircraft change delays. The occurrence probability of an event e is denoted by P^e and the actual delay (the elapsed time) of event e is denoted by $\hat{\phi}^e$, which forms a probability density function, $\Phi^e(\phi)$. The time when an event may occur is also uncertain; some events may occur early and allow airline ground staff sufficient time to respond, while some may tend to occur at a later stage, e.g. delays due to the engineering check, and leave airlines little time to respond but delaying a flight. The time event e occurs is, hence modelled as a stochastic variable $\hat{\phi}_s^e$, which forms a probability density function, $\Phi_s^e(\phi)$. Hence, the "realised time" of a disrupting event during the turnaround operation of flight f_{ij} is modelled by (3.14). Together with (3.12) and (3.13) given earlier, we are able to model the actual departure time of flight f_{ij} and its corresponding departure delay.

$$\hat{t}_i^{events} = \hat{\phi}_s^e + \hat{\phi}^e \tag{3.14}$$

3.4.4 Model Implementation – Model Parameters and Flight Data

The Turnaround Simulation (TS) model was described by the set of equations from (3.12) to (3.14). Given the complex combination of semi-Markov Chains and discrete event modelling, the Monte Carlo simulation technique was employed to implement the TS model for case studies. The Monte Carlo simulation is a technique widely used in modelling complex systems, especially those systems that are composed of stochastic sub-systems or stochastic components. A major advantage of using Monte Carlo simulation techniques is that modellers are able to change model parameters and the stochastic behaviour of system components in order to test "what-if" scenario study questions. This feature is essential when we study a complex system, as we are often interested in understanding how the system may behave under different system control strategies. Two examples include: (1) the modelling of the National Air Space (NAS) in the U.S. (NASA 2008); and (2) the MEANS model by Clarke et al. (2007) which modelled the dynamics of airline scheduling and air traffic management strategies. Readers who would like to understand more about this technique are encouraged to consult Fishman (1997) and other relevant texts.

Model parameters in the TS model were compiled from airline data, including the occurrence probabilities of disrupting events, service time of various turnaround activities, and their corresponding probability distributions. These parameters were used to simulate the major components of the TS model, including the semi-

Markov Chain and discrete event simulation. A full turnaround operation was simulated for 1,000 times in order to control simulation "noises" (Wu 2006) and meet statistical sampling requirements. Simulation results were then compared with real airline operational data in order to validate the TS model and to calibrate model parameters, following the standard simulation model building procedures (Klafehn et al. 1996).

To implement the TS model and conduct case studies, the same set of flight data from Airline R used previously in Chapter 2 was used here. Two typical European city shuttle services, denoted by flight RR-X and RR-Y were used in the following case studies. RR-X was scheduled to depart at 19.45 hours with a turnaround time of 60 minutes. Historical data of RR-X revealed that Beta (4,9) was statistically suitable for modelling the arrival distribution of inbound flights; Beta (2,5) was statistically sound for RR-Y, on the other hand. Moreover, the use of the same set of data in two different models, i.e. the TTA model in Chapter 2 and the TS model here, creates an opportunity for exploring the perspectives that different modelling methodologies may bring to the same operational problem.

Model parameters required to implement the TS model include state sojourn time functions of various states, i.e. the service time probability functions of activities and a state transition probability matrix, representing the probability of transition from one state to another. Transition probabilities between states in the workflow of cargo and baggage processing are given in Table 3.12. State sojourn time functions used in the simulation include the Normal function, Beta function and Exponential function. These parameters were generated by statistical analyses of historical flight data. Due to the lack of detailed delay codes in the data obtained from Airline R, the following parameters were generated for model demonstration purposes with the contribution of expert judgements from Airline R. More precise model parameters can be obtained by analysing historical flight data, once more detailed data are available. From Table 3.12, we can see that the probability of incurring delays due to cargo/baggage loading was 0.1 and the probability of incurring baggage loading disruption due to late check-in of bags was 0.02 for Airline R's operations.

The transition probability matrix for passenger and cabin cleaning is provided in Table 3.13. As seen, the operation of this workflow tended to incur delays and disruptions due to passengers, either because of check-in issues or because of missing checked-in passengers in terminals. The possibility of incurring delays for these reasons to the departure procedures before pushing back an aircraft was relatively low, by 0.001.

Four major disrupting events, namely fuelling delay, engineering check delay, aircraft damage on the ramp, and un-scheduled aircraft change on the day of operation were included to model the occurrence of independent disrupting events during aircraft turnaround operations. The parameters used in modelling discrete events are given below in Table 3.14. We found that delays due to fuelling and engineering check were not uncommon for this airline. The occurrence time of disruption due to fuelling was rather early (Exponential(10) distribution), while disruptions due to engineering checks tended to occur at a later stage (Exponential(30) distribution), causing a high likelihood of incurring departure delays. Although the probability of incurring disruptions due to aircraft damage and last-minute aircraft change was rather low, the potential impacts (delays) were significant, averaging around 30 to 45 minutes for aircraft damage and aircraft change respectively.

Table 3.12 State transition probability in cargo and baggage processing

States	1[†]	2	3	4	5	6	7	8	9
1	0.0/B	1.0	–	–	–	–	–	–	–
2	–	0.0/N	0.90	–	0.05	0.05	–	–	–
3	–	–	0.0/N	0.80	–	–	0.1	0.08	0.02
4	–	–	–	1.0/B	–	–	–	–	–
5	–	1.0	–	–	0.0/E	–	–	–	–
6	–	1.0	–	–	–	0.0/E	–	–	–
7	–	–	1.0	–	–	–	0.0/E	–	–
8	–	–	1.0	–	–	–	–	0.0/E	–
9	–	–	1.0	–	–	–	–	–	0.0/E

Note: [†]State sojourn time function: B (Beta), E (Exponential), and N (Normal).

Table 3.13 State transition probability in passenger/crew/cabin flow

States	1[†]	2	3	4	5	6	7	8	9	10	11	12	13
1	0.0/B	1.0	-	-	-	-	-	-	-	-	-	-	-
2	-	0.0/N	1.0	-	-	-	-	-	-	-	-	-	-
3	-	-	0.0/N	1.0	-	-	-	-	-	-	-	-	-
4	-	-	-	0.0	1.0	-	-	-	-	-	-	-	-
5	-	-	-	-	0.0/N	0.80	-	0.02	0.10	0.08	-	-	-
6	-	-	-	-	-	0.0/N	0.95	-	-	-	0.019	0.03	0.001
7	-	-	-	-	-	-	1.0/B	-	-	-	-	-	-
8	-	-	-	-	1.0	-	-	0.0/E	-	-	-	-	-
9	-	-	-	-	1.0	-	-	-	0.0/E	-	-	-	-
10	-	-	-	-	1.0	-	-	-	-	0.0/E	-	-	-
11	-	-	-	-	-	1.0	-	-	-	-	0.0/E	-	-
12	-	-	-	-	-	1.0	-	-	-	-	-	0.0/E	-
13	-	-	-	-	-	1.0	-	-	-	-	-	-	0.0/E

Note: [†]State sojourn time function: B (Beta), E (Exponential), and N (Normal).

Table 3.14 Disrupting events in aircraft turnaround operations

Event	Event Description	Occurrence Probability	Occurrence Epoch†	Event Duration
1	Fuelling Activity Delay	0.02	Exponential (10)	Normal (15,3)
2	Engineering Check Delay	0.02	Exponential (30)	Normal (20,5)
3	Aircraft Damage	0.005	Exponential (15)	Normal (30,5)
4	Aircraft Changes	0.002	Exponential (15)	Normal (45,5)

Note: †The time from the start of aircraft turnaround operations.

3.4.5 Case Study Results

The TS model was implemented with the previously given model parameters. The simulated departure punctuality of RR-X and RR-Y is illustrated in Figure 3.18 and is compared with the observed punctuality records. The departure punctuality of study flights was expressed as cumulative density functions (CDFs) and it is seen in Figure 3.18 that simulation results of RR-X from the TS model matched closely with the observed punctuality data. However, the simulated departure punctuality of RR-Y from the TS model did not quite match the observed departure punctuality, which showed rather poor on-time performance during the operation of RR-Y.

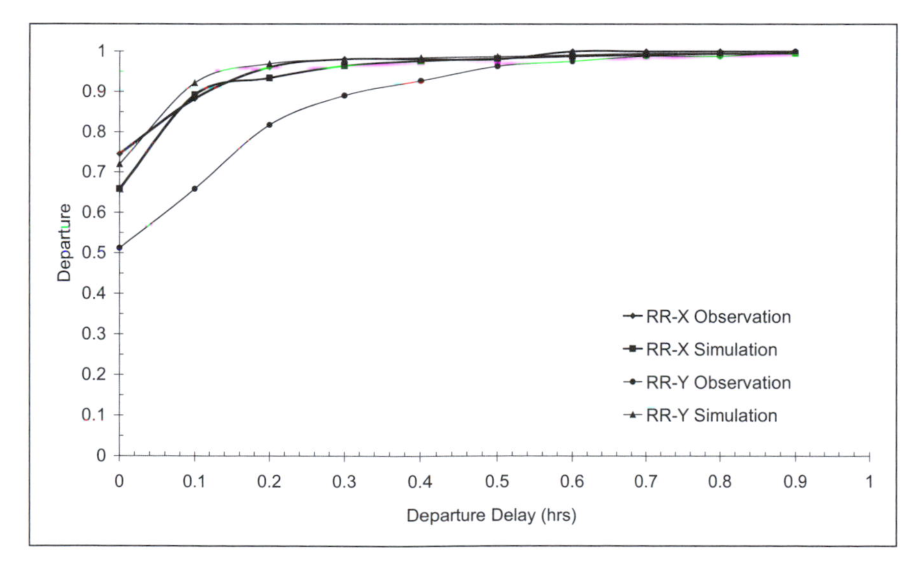

Figure 3.18 Observed and simulated on-time performance of case study flights

The scheduled ground time of RR-Y was 65 minutes, so some 20 minutes of schedule buffer time was included in the flight schedule (if the required standard turnaround time for a B757 European operation was 45 minutes). It was seen from the detailed simulation results given in Table 3.15 that RR-Y should be able to deliver a punctual service because of a longer ground time than RR-X. According to simulation results in Table 3.15, RR-Y's departure on-time performance should be as good as RR-X's as seen in Figure 3.18. Since the arrival punctuality of RR-Y from observations was 58 per cent, it was hence speculated that the observed poor departure punctuality of RR-Y was caused by poor operational efficiency and unexpected disruptions in the turnaround of RR-Y.

The decrease of operational efficiency of aircraft turnaround may result from labour shortage, equipment availability for ground services, and airport capacity constraints. The first two situations are more often experienced during the peak hours of airport operations, and the airport capacity constraints come from terminal capacity and runway/taxiway capacity. For instance, gate allocation for inbound aircraft can be disrupted due to the extended occupancy time of delayed flights, and runway/taxiway congestion may delay aircraft push back and taxiing. To study these potential causes of poor departure OTP of RR-Y, two scenario analyses were carried out to model RR-Y in different operational conditions. Scenario A simulated RR-Y in a condition of low aircraft turnaround efficiency. Scenario B simulated the same situation as Scenario A together with airport ground congestion for departures. An average departure push-back delay of two minutes due to airport ground congestion was included in Scenario B. Results of scenario studies are given in Table 3.15 and illustrated in Figure 3.19.

Table 3.15 Simulation results of turnaround operations of study flights

	Inbound Delay[†]	Turnaround Time[*]	Operational Delay	Outbound Delay
RR-X	2.3	51	2.4	2.6
RR-Y	2.7	51	1.9	2.2
RR-Y in Scenario A	2.7	61	4.2	4.5
RR-Y in Scenario B	2.7	61	4.2	6.3
RR-Y in Scenario C	2.7	67	7.1	7.4
RR-Y in Scenario D	2.7	71	9.6	9.8
RR-Y in Scenario E	2.7	75	13.1	13.3

Notes: [†]Mean time (minutes) of simulation flights and [*]Mean service time of simulation flights.

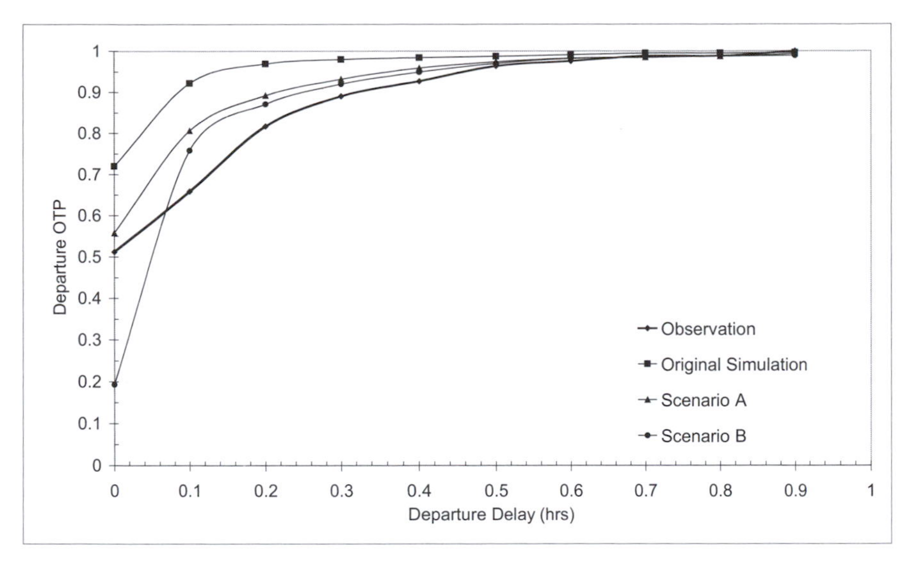

Figure 3.19 Scenario analysis of RR-Y

Statistics in Table 3.15 shows that the mean service time of RR-Y increased from 51 minutes in the original case to 61 minutes in Scenario A, due to the low turnaround efficiency, i.e. a longer operational time required for ground services. The mean outbound delay of RR-Y in Scenario A consequently increased to 4.5 minutes. In Scenario B, the mean service time remained 61 minutes, but the mean outbound delay increased to 6.3 minutes, because of the increase of airport congestion. When the simulated departure punctuality of RR-Y was compared with observations in Figure 3.19, it is seen that RR-Y was more likely to suffer from poor punctuality due to low turnaround efficiency. On the other hand, simulation results showed that delays from airport ground congestion only contributed a relatively small portion to departure delays in Scenario B (an extra outbound delay of 1.8 minutes on average) and did not affect the overall departure delay curve of RR-Y.

To further investigate, three more scenario studies of RR-Y in the situation of low turnaround efficiency were conducted. Scenario C modelled a turnaround operation that required 55 minutes to finish all aircraft ground services (the standard time was 45 minutes). Scenario D and E modelled the same condition but required 60 and 65 minutes of turnaround service time respectively. In other words, we are investigating the likelihood of lower ground service efficiency on the departure punctuality of RR-Y.

Simulation results from Scenario C, D and E are shown in Figure 3.20. It is seen that the observed departure punctuality of RR-Y matched closely with the departure CDF of Scenario C, which consumed 67 minutes on average to finish all turnaround services for RR-Y. Hence, it was suggested by simulation results that low turnaround efficiency of RR-Y could be the major cause of poor punctuality. Unfortunately, detailed delay codes of RR-Y were not available from the data resource to validate the speculation that low turnaround efficiency resulted in the poor departure punctuality of RR-Y. However, simulation results provided some evidence to support our hypothesis for flight RR-Y.

3.4.6 Implications for Airline Scheduling and Operations

It was found from the case study of RR-X that if the efficiency of aircraft turnaround operations at an airport was assumed to be consistent, the "inherent" schedule punctuality, which reflected the turnaround efficiency and the amount of scheduled buffer time for ground operations, could be estimated before the implementation of flight schedules. For instance, RR-X was scheduled with 15-minute schedule buffer time when using a B757 aircraft, so the expected punctuality of RR-X can be approximated by the punctuality curves from simulations shown earlier in Figure 3.18. However, the observed punctuality of RR-Y showed that operational variance existed in turnaround operations and may consequently influence flight punctuality. We will come back to the topic of "inherent delays/punctuality" in Chapter 5 with a more in-depth discussion.

The major advantage of the TS model introduced in this chapter, when compared with the PERT model, is that the TS model is able to simulate the stochastic and dynamic transition behaviour between ground service activities as well as to model the occurrence of disruptions to aircraft turnaround. The occurrence probability and duration of operational disruptions can be obtained from historical flight data.

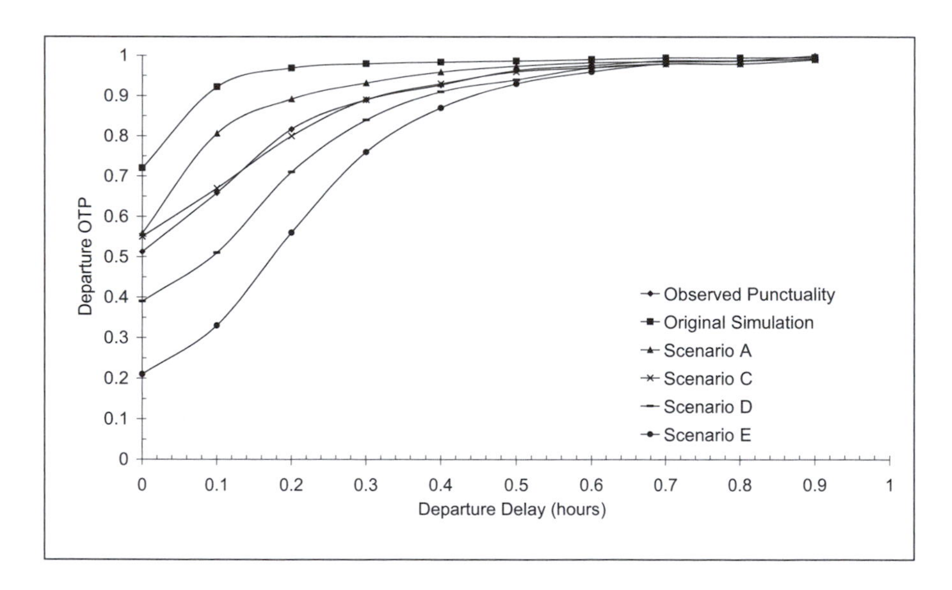

Figure 3.20 Low turnaround efficiency scenario analysis for RR-Y

These statistics also inform an airline of how often a specific disrupting event has occurred in the past, and when this disruption was more likely to occur. Operational improvements can be developed and implemented, targeting at improving those "weak links" in airline ground operations.

Compared with the TTA model introduced in Chapter 2, the advantage of the TS model is that it is able to simulate on a micro level the operations of individual activities and the occurrence of disruptions. Although the stochastic features of airline operations were captured in the TTA model by using stochastic flight arrival and departure functions, the TTA model was an aggregate model that had its strength in studying strategic issues such as the allocation of turnaround buffer time for different flights at different airports. In this regard, the TS model is quite suitable for conducting micro scenario analyses such as investigating the impact of improved ground service efficiency (reducing turnaround service time) or the impact of unexpected disruptions on airline operations.

For OTP benchmarking purposes, both TTA and TS models are able to provide benchmarks before schedule implementation. The concept of "inherent delays" of a flight schedule originates from the development of OTP benchmarking measures, and has important implications for airline scheduling and operational benchmarking both before and after schedule executions (Wu 2006). This concept will be further discussed in detail in Chapter 5.

3.5 Managing Passenger Flows at Airports

3.5.1 Passenger Flows – from Check-in/Transit to Boarding

A major task of airline ground operations at an airport is to manage passengers and the resultant "flows" of passengers from check-in, immigration screening (only for international operations), security check, transit services (only for transit passengers), to boarding an aircraft at a gate. These series of operations are operated by various agencies including: airlines, immigration agencies, and security agencies, and are facilitated by an airport authority which provides the infrastructure, i.e. airport terminals and facilities. From the operational perspective of an airline, the goal of managing passenger flows at an airport is to provide quality travel services to air passengers. Airlines usually only control a few activities in passenger flow management, namely passenger check-in and passenger boarding at gates, although some airlines are more actively involved in facilitating passenger immigration and security screening, especially for premium passengers. Apart from facilitating passenger flows and providing high service quality to customers, an important operational goal for airlines is to ensure a smooth process for passengers from check-in to boarding and ensuring there are no delays to flight departures due to passenger handling at airports.

Airports play an important role in facilitating and managing passenger flows at terminals. First and foremost, an airport provides airlines with facilities (e.g.

check-in counters) and terminals that are designed for efficiently managing passenger flows. Second, an airport operator also provides (or facilitates) some relevant services such as immigration screening (for international operations) and security check for passengers. The capacity of these services and the capacity of airport terminals are critical in managing passenger movement in an airport; insufficient capacity often causes terminal congestion, passenger delays as well as a low level of service quality. From an airport operator's perspective, speedy processing of passengers from check-in to the completion of security checks also brings potential financial benefits to an airport, because passengers will have more free time to spend on "airport shopping", or at other retail businesses provided at airport terminals (Francis et al. 2003).

3.5.2 Passenger Arrival Patterns at Check-in Counters

The arrival patterns of air passengers at an airport are highly related to the scheduling of flights (departure times) by airlines at a specific airport. The "access time" of an airport is largely stochastic for a passenger, considering the uncertainties involved in the travel from home to the airport by ground transport modes (Kim et al. 2004; Ndoh and Ashford 1993). Moreover, expecting a series of processes from check-in to boarding, most passengers arrive at the airport early to ensure that they can catch their flights in time. Hence, the arrival of passengers at the check-in lounge of an airport depends upon passenger expectation (and/or perception) of airport access time, time required to finish the processes before boarding, and the departure time of a scheduled flight. During peak hours, more passengers arrive than in off-peak hours, because more flights are scheduled to depart during peak hours. However, the "peak" of passenger arrivals for check-in is always earlier than the scheduled peak departure time at an airport.

The arrival pattern of passengers corresponding to a scheduled flight has been well studied by internal projects of airlines and airports for reducing check-in queues by balancing the opening of check-in counters, staffing at counters, and the expected arrival patterns of passengers (e.g. Kim et al. 2004; Park and Ahn 2003). The actual arrival pattern of passengers is often obtained by on-site observations or through passenger survey at airports. Figure 3.21 illustrates a common arrival pattern of passengers at check-in counters of an airport. This pattern resembles the one provided by Park and Ahn (2003) from a passenger survey at Seoul Gimpo International Airport in Korea. It is noted that the "shape" of the arrival pattern is not necessarily the same for all flights, because a large aircraft naturally carries more passengers than a small one. Hence, more check-in counters are needed for a flight operated by a large aircraft and counters are also opened earlier in order to accommodate the volume of check-in passengers.

Passengers start arriving at the check-in lounge about two to three hours before the scheduled departure time of a flight. Airlines use the passenger arrival pattern to allocate resources for passenger check-in, including human resources (check-in staff) and physical resources (check-in counters). At the early stage of passenger

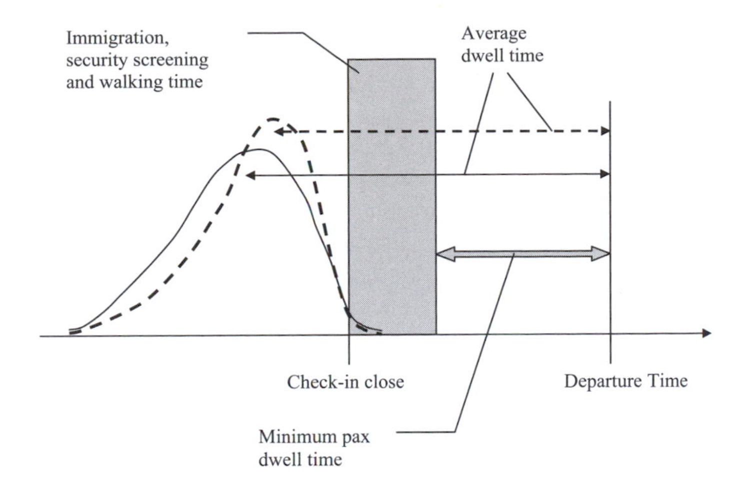

Figure 3.21 Passenger arrival patterns and the corresponding passenger dwell time

arrival, only limited counters and staff are needed. As the volume of passenger arrivals increases, an airline will open more counters and allocate more resources for the passenger check-in service, and through this, maintain a certain level of service quality by managing waiting time of passengers in check-in queues. Both the human resources and check-in counters cost airline money. Hence, a good prediction of passenger arrival patterns and good management of resource allocation for passenger check-in will reduce the operating costs of an airline and improve the financial bottom line as well.

While much attention has been paid to modelling and observing passenger arrival patterns, what has been overlooked in the past is that airlines' own operations may also cause changes to passenger arrival patterns at an airport. For instance, the recent advances in technology bring passengers more check-in options such as online check-in services. Air passengers can now check in up to 48 hours before the flight departure time by various means, e.g. telephone, fax or internet. To encourage passengers to check in before they arrive at airports, airlines often offer "express" check-in queues or check-in kiosks at airports for those passengers to drop off bags. These new check-in options reduce the demand for check-in counters at airports (resulting in lower operating costs for airlines) and meanwhile, provide passengers with flexibility for check-in. However, the implications of this operational advance by airlines go far beyond the issues of providing more check-in options for passengers and reducing operating costs associated with passenger check-in at airports. These airline operational improvement measures result in potential changes in the arrival patterns of passengers at an airport.

Some airlines have already observed the trend that internet checked-in passengers tend to arrive later than before. In some domestic operations, passengers can even print boarding passes (with barcodes) by checking in on the internet, then bypass the check-in lounge and go straight to security screening for departure (if they have no check-in baggage). The implication of this change is that: first, passenger "dwell time" at an airport can be significantly reduced, especially for domestic operations; second, the arrival peaks at the services following check-in, e.g. immigration and security screening, can be higher than before due to the increased concentration of passenger arrivals.

The definition of *dwell time* is: *the time between a passenger's arrival at a check-in lounge and the scheduled departure time of the flight that the passenger checks in for*. Some portions of dwell time are unavoidable and must be spent at certain processes such as check-in queues, immigration checks (for international flights), security screening, and even walking in the terminal. The remaining of the dwell time is defined as the *free dwell time* that is available for a passenger to conduct other activities such as shopping, eating, or resting in the terminal area, waiting for boarding an aircraft. It can be seen from Figure 3.21 that the later a passenger arrives, the less free dwell time a passenger will enjoy. Also, the shorter the time required to finish check-in, immigration screen, security check and walking, the longer the free dwell time available to a passenger.

In such a situation, airlines may gain in terms of operating cost reduction and better service quality to passengers due to less waiting time spent at check-in counters or even no on-site check-in at all. However, from an airport operator's perspective, the reduction of passenger dwell time at an airport and consequently the reduction of free dwell time may cause negative impacts on airport retail businesses, if the traditional wisdom and empirical observation that "the more free dwell time a passenger has, the more likely a passenger will spend" is true for most passengers (Papatheodorou and Lei 2006; Torres et al. 2005). In Addition, the shift of passenger arrival patterns at an airport may cause pressure on some services such as immigration and security check that are already under pressure due to increased security measures adopted by airports and aviation authorities worldwide after the 9/11 attack in 2001. When passengers spend more time in these time-consuming procedures at an airport, passengers will have less free dwell time and higher levels of emotional stress. This may cause a potential reduction in retail revenues for airports. This effect has also been observed by Francis et al. (2003) in a study of a secondary airport in Europe.

3.5.3 Passenger Behaviour at Terminals and Airport Operations

Passenger "behaviour" at a terminal has two dimensions, namely spatial behaviour and temporal behaviour. The spatial behaviour of a passenger describes the physical relationship between the location of a passenger and the environment, i.e. airport terminals. A well designed terminal guides passengers through various "decision points", e.g. from a check-in lounge to immigration check, by using various tools including signs and the building itself by means such as a corridor linking two processes in a terminal. Passenger way-finding studies have evolved from physically quantifying way-finding (the spatial behaviour) by a number of factors, e.g. size of building and number of decision points, to involving some human elements less thought about such as the spatial behaviour of passenger in a terminal and the visibility of signs to passengers (Churchill et al. 2008). The use of a visibility index (VI), proposed originally by Braaksma and Cook (1980), was further improved to consider the weights of different visual contacts in calculating VI by Tosic and Basic (1984). Often, passengers are not aware of the full layout of terminals in an airport, except for highly experienced and frequent travellers. Hence, the spatial behaviour of a passenger in an airport is often guided or even "induced" by a simplified "corridor layout" in a terminal created by the airport operator.

The temporal dimension of passenger behaviour in an airport is more associated with the "time" a passenger may make use of during the dwell time. The previously defined dwell time and free dwell time of a passenger in an airport describes the temporal element of a passenger's movement from arriving at the check-in lounge to boarding an aircraft at the designated gate. Obviously, the temporal dimension of passenger movement is influenced by the series of processes a passenger needs to go through as well as being influenced by uncertainties involved in those processes such as service queues and delays. The temporal behaviour of passengers is of concern to an airline because flight delays may be caused by a missing checked-in passenger who is not aware of the flight departure time or even not aware of the time required to travel from his/her current location to the boarding gate.

The combined effect of the temporal and spatial behaviour of passengers in a terminal is of special interest to the airport operator. Airport revenues come from two streams, namely aeronautical revenues and non-aeronautical revenues. Aeronautical revenues of an airport mainly come from the operation of aeronautical activities, i.e. transporting passengers and freights by airlines. A major contributor to aeronautical revenues is landing fees coming from airlines. On the other hand, non-aeronautical revenues are sourced from retail sales, airport property development and commercial activities, e.g. car parks and rental car operations.

Retail businesses play a crucial role in the overall financial portfolio of an airport business. The contribution of retail revenue to an airport can range from 23 per cent (AUS$174 million) of the total revenue of Sydney Airport in 2007, 22 per cent (£241 million) of London Heathrow Airport in 2006, to 26 per cent (€301 million) of Amsterdam Airport Schiphol (BAA 2006; Schiphol Group 2007; Sydney Airport 2007). This revenue stream is crucial for improving the financial performance of an airport business. Together with separate rental revenues generated from retail spaces, the combined non-aeronautical revenue to an airport can be as high as 40 per cent of the total annual revenue. Since the sales targets of airport retailing are mostly air passengers, the combined spatial and temporal behaviour of passengers significantly influences the consumption behaviour of passengers at an airport.

In general, those factors that may influence passengers' behaviour (spatial and temporal) at an airport can be classified into two categories: exogenous and endogenous factors to the passengers. Exogenously, airline operations, airport operations and airport layout design influence passengers' behaviour. The aforementioned innovations in passenger handling technologies such as the internet check-in service, may change the arrival pattern of passengers and consequently influence the dwell time a passenger may have at the airport. Airport operations, in particular the facilitation of immigration screening and security check affect the "free" dwell time a passenger may have. Accordingly, this time constraint of a passenger's spatial behaviour may likewise constrain potential consumptions by a passenger during his/her free dwell time in a terminal, impacting the retail business of an airport. Airport layout also exogenously influences a passenger's spatial and temporal behaviour. The more time that is spent on way-finding or walking, the less free time a passenger will have to conduct other activities including shopping at an airport.

On the other hand, a passenger's spatial and temporal behaviour is profoundly influenced by his/her own socio-demographic characteristics such as income level, consumption preferences, education level and even travel purposes. Hence, how well in advance a passenger may arrive at an airport is indeed a "personal behaviour and preference" issue. These types of travel preferences of passengers have been research topics not only in aviation but also in other fields such as marketing, and transport choice studies (Hensher et al. 2005). Although exogenous factors may limit a passenger's spatial and temporal behaviour, these behaviours are also subject to the influence of the passenger's own preferences, in particular the consumption preferences in the airport terminal environment.

A study by Appold and Kasarda (2006) on the scale of U.S. airport retail activities showed that both exogenous and endogenous factors influence passengers' behaviour in an airport terminal, in particular their consumption behaviour. Some retailers tend to attract a high volume of passengers (e.g. a food store), while others have little by way of transaction volume but high unit sale values (e.g. a branded good). Clearly, the free dwell time of a passenger is an important factor (though in some cases, this factor may not be statistically significant), but more profoundly those endogenous factors, i.e. passengers' consumption preferences determine passengers' consumption behaviour in a terminal.

3.5.4 Managing Passenger Boarding at Gates

Managing passenger boarding at a gate is an important operation which for many cases also lies on the "critical path" of aircraft turnaround operations, as pointed out earlier in this chapter. Since passenger boarding is one of the last activities to conduct before preparing an aircraft for departure, this activity has a high potential to cause departure delays. Within the processes from passenger check-in to passenger boarding at an airport, airlines only have control over the first and the last activity. In many delay cases due to late passenger boarding, delays occur to passengers on the path from the check-in lounge to the boarding gate, instead of boarding itself.

Delays to passenger boarding could be due to late check-in passengers, although more cases are due to missing checked-in passengers in the terminal. Although airlines often require passengers to arrive at the assigned boarding gate by a certain time prior to departure, facing a late or missing checked-in passenger is always a dilemma for an airline. Since the passenger has already checked in, it takes time to retrieve the checked luggage from the cargo hold of an aircraft. For a jumbo jet, it may take as long as 20 minutes to retrieve a bag from the loaded cargo hold, causing delays to a departure flight. Alternatively, airline ground staff will try to locate the missing passenger by searching in retail shops and via radio broadcasts in the terminal. Either of these cases have a chance of causing delays to a flight.

Apart from delays due to passengers, delays also occur with the passenger boarding process itself. As discussed earlier in Section 3.3, passenger boarding is a time-consuming process, especially for the turnaround of a large aircraft, e.g. A380 or B747. Technically, the more passengers to board an aircraft, the longer it takes to finish passenger boarding. In the industry, airlines employ different passenger boarding methods; some airlines follow the "conventional boarding method" (i.e. boarding from the back to the front rows of an aircraft), while others create more advanced but complex boarding methods in order to reduce boarding time and hopefully reduce aircraft turnaround time. A cited ground (opportunity) cost by a U.S. carrier in Nyquist and McFadden (2008) was US$30 per minute on the ground. This low unit cost can easily add up to a formidable amount for a large carrier across a network; this is why shortening aircraft turnaround time and improving aircraft utilisation is so critical for airline profitability.

The conventional passenger boarding method widely used in the industry is called *Back-to-Front Boarding*, i.e. boarding passengers from the back of a plane by a section of rows each time to the front of a plane. In this conventional boarding method, airlines often board first/business passengers and those passengers who need assistance in boarding before starting general boarding of the remaining passengers. The principle of using the Back-to-Front Boarding method is to allow passengers space once aboard in order to efficiently secure carry-on luggage as well as locate a seat. However, this boarding method is not completely efficient, as obviously when rear section passengers are boarding, the front section of the plane is not used. Hence, various computer simulation models were run in search of the "optimal" passenger boarding method in the shortest time.

In two recently published papers by Nyquist and McFadden (2008) and Steffen (2008), various boarding methods were tested via computer simulation models. Results showed that the conventional Back-to-Front method was not the optimal choice in terms of boarding time, and the *Front-to-Back* method was widely known as the worst boarding method (i.e. *the worst case*), because boarding passengers in the front section first blocks all remaining passengers to other sections of the plane. By using a more complex boarding sequence, the optimal boarding method, called the *Steffen* model, could potentially save up to 79 per cent of boarding time compared to the *worst case* (Steffen 2008). The conventional method, however, saves only 25 per cent boarding time when compared with the worst case. Since the *Steffen* model is too complex to implement and not practically feasible, a "modified optimal" boarding method based on the optimal model was tested; this modified optimal boarding model could still save up to 58 per cent of boarding time relative to the worst case.

Improvements on boarding time saving mainly come from boarding passengers in such a sequence that combines passenger boarding from the back to the front of a plane, separate odd/even row boarding, and meanwhile window-seat passengers board first, then middle-seat and aisle-seat passengers. The efficiency gain of this "modified optimal" method comes from efficient use of the whole cabin space for passenger boarding simultaneously, although some inconveniences may occur in practice, e.g. people travelling in groups will not board together under such a scheme. However, in reality, travellers are more likely to ignore such a complex boarding scheme and still board together. This scheme is similar to the current one used by Southwest Airlines in the U.S., called the *Open Seating* method. The Open Seating method assigns passengers a group number and a boarding number based on the time when a passenger checks in. When boarding starts, groups are called in a specific order and passengers in different boarding groups line up according to individual boarding numbers. Passengers have the opportunity to choose their own seats once aboard.

A slight variation of the Open Seating model and the Steffen model is the so-called *Reverse Pyramid* boarding model which was developed by the then America West Airlines (see Van den Briel et al. (2005) for more details). This scheme delicately combines the concepts of separate row boarding, back-to-front and window-middle-aisle seat boarding. The Reverse Pyramid model is illustrated in Figure 3.22. The numbers of seat blocks represent the sequence of boarding. We can see that the boarding sequence starts from the window seats of the rear section of the cabin, then proceeds to board the middle seats of the rear section and simultaneously, boards the window seats of the middle section of the cabin. The Reverse Pyramid model can save boarding time due to the reduction of potential passenger boarding conflicts between rows and between seats, as well as by using the whole cabin space for boarding simultaneously.

For some low-cost carriers who do not assign seats to passengers (i.e. free seating), the *Random Boarding* method is widely used. Passengers board a plane randomly without a pre-assigned sequence and can choose any seats aboard based on a first-come-first-serve basis. Although seemingly a "chaotic" boarding procedure without specific sequences, this boarding method outperforms the conventional Back-to-Front method with a time saving of 40 per cent according to the simulation by Steffen (2008). This is why this boarding method has been widely used by low-cost carriers due to the fast boarding time and the simplicity (also cost reduction) in implementation.

It should be noted that most simulation models on passenger boarding involve certain model assumptions and simplification including the queueing behaviour of passengers, non-group boarding, the amount of carry-on luggage and the time to store

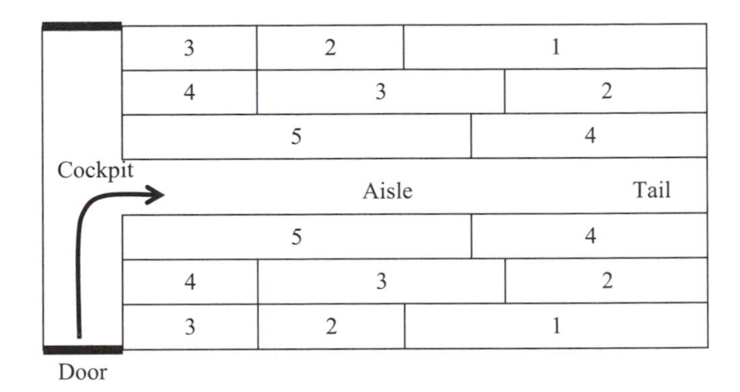

Figure 3.22 Reverse pyramid boarding model

carry-on luggage. These assumptions greatly reduce the complexity in modelling passenger boarding and hence, simulation results are often more optimistic than real-world operations by airlines. The random behaviour of passenger arrival at a gate also contributes to the uncertainties involved in boarding passengers. Passengers are often distracted, largely by retail shops in a terminal, from reaching the assigned arrival gate in the shortest time. While this "distraction" is essential for airport retail businesses and airport finance, it directly conflicts with passenger processing in airline operations at an airport. To resolve this conflict and improve the timeliness of passenger boarding, most airport operators have agreements in place with retailers to prevent them from selling products to a passenger whose flight departs within a certain time, e.g. in 30 minutes.

3.6 Summary

In Chapter 3 we have discussed some of the issues and challenges of airline ground operations at an airport including resource connections, stochastic disruptions, and the airline's own scheduling constraints. A PERT model was built to manage aircraft turnaround operations which included many activities such as passenger handling, the baggage/cargo processing, engineering checks and catering services. Real examples were given to demonstrate how the PERT model could be used in the industry to manage aircraft turnaround operations. A Semi-Markov Chain model (namely the TS model) was presented to describe, on a micro level, those activities within aircraft turnaround operations as well as to model the stochastic disruptions an airline faces in daily ground operations.

The same set of flight data used in Chapter 2 was used again here to demonstrate the differences between the macro model (TTA model in Chapter 2) and the micro model (TS model in this chapter) in modelling airline operations and the stochastic factors involved. Case study results showed the advantages of the TS model in modelling individual and stochastic activities in aircraft ground operations. Finally, Chapter 3 also discussed strategies on managing passenger flows in an airport environment from both an airport operator's and an airline's perspective. Managerial and operational implications from managing passenger flows were discussed, focusing on the implications for airport retail revenues and airline operations.

Until now we have discussed airline operations on the ground in Chapters 2 and 3. In the following Chapter 4, we will start introducing the readers to airline operations on a network scale. The "network effects" of airline operations will be further explored by considering the complexity of airline operations, airline scheduling, and the operational impacts of individual flights on an airline network. Unlike previous chapters, we will approach airline scheduling and operational issues from a "network perspective" and explain how modern airlines manage gigantic flight networks as well as millions of passengers in daily operations.

Appendix
Notations and Symbols Introduced in Chapter 3

ES_j	the earliest start time of an activity *j* (at a node *j*) in a PERT network
EC_i	the earliest completion time of an activity *i* in a PERT network
LC_j	the latest completion time of activity *j* in a PERT network
LS_j	the latest start time of activity *j* in a PERT network
D_j	the duration of activity *j* in a PERT network
F	the set of all flights in a schedule
$\hat{t}_i^{cargo}$	the realised (actual) time to finish cargo processing of f_{ij}
$\hat{t}_i^{pax}$	the realised (actual) time to finish passenger processing of f_{ij}
$\hat{t}_i^{events}$	the realised (actual) time to a disrupting event during turnaround of f_{ij}
K	the set of service activities of cargo and baggage workflow
P_{pq}	the transition probability from state p to state q
$\hat{\phi}_{pq}^{k}$	the realised (actual) time to finish a cargo activity k (in state p) before transiting to state q in the cargo workflow
$\Phi_{pq}^{k}(\phi)$	the probability density function of $\hat{\phi}_{pq}^{k}$
Ω	the set of service activities of passenger processing workflow
$\hat{\phi}_{pq}^{\omega}$	the realised (actual) time to finish a passenger processing activity ω (in state p) before transiting to state q in the passenger workflow
$\Phi_{pq}^{\omega}(\phi)$	the probability density function of $\hat{\phi}_{pq}^{\omega}$

P^e	the probability that a disrupting event may occur
$\hat{\phi}^e$	the realised (actual) delay due to event e
$\Phi^e(\phi)$	the probability density function of $\hat{\phi}^e$
$\hat{\phi}_s^e$	the realised (actual) time when event e occurs
$\Phi_s^e(\phi)$	the probability density function of $\hat{\phi}_s^e$

Chapter 4
Enroute Flight Operations in Airline Networks[1]

The discussion in earlier chapters was focused on airline operations at airports. This chapter extends the discussion of airline operations to a network level with the introduction of enroute flight operations by airlines, focusing on two main areas, namely crewing and fleeting. Key issues of crewing and fleeting are discussed in Sections 4.1 and 4.2, including setting up crew bases, regulation limits of working hours, aircraft maintenance requirements and routing aircraft for maintenance needs in a network. The complexity emerging from the synchronisation between flights, fleets, crew and passenger itineraries is discussed in Section 4.3 from the perspective of airline operations.

Section 4.3 aims to address the operational problems of a complex airline network, starting with exploring the complexity in planning airline networks, operational problems emerging from such networks, and how airline network complexity is measured empirically both at the stage of schedule planning and the stage of post-operation analysis as seen in Section 4.4. The issue of delay propagation in an airline network is discussed in the last section of this chapter. Mathematical models are introduced to describe the mechanism of delay propagation between flights among sub-networks of fleeting, crewing and aircraft routing. Managerial and operational implications of delay propagation are examined in depth from the viewpoint of airline scheduling and operational management, shedding some light on how to improve airline operational reliability.

4.1 Fleets and Aircraft Routing

Previously in Chapters 2 and 3, the discussion was primarily focused on airline operations at airports including aircraft turnaround operations, passenger flows, goods/cargo/baggage processing, and the operational constraints imposed by airline schedules and the operating environment. In the sense of an airline network, in which airports are modelled as "nodes" and flights between nodes are modelled as "arcs" (or "links"), we are now going to examine airline operations, moving from a previous "node perspective" to a linked and dynamic "network perspective". To demonstrate how an airline network is formed, we will introduce two major operations that are performed on a network scale, namely aircraft routing and crew

1 This chapter is partially based on the following publications: Wu (2005); (2006).

scheduling, and explain how those "arcs" in the network map of an airline are planned.

4.1.1 Fleet Assignment and Demand Uncertainties

Fleet assignment plays a critical role in the overall commercial and operational performance of an airline schedule. The primary task of fleet assignment and hence the various and complex fleet assignment models is to allocate available airline resources (in particular, aircraft capacity) to best match uncertain passenger demands for flights due to airline competition and the local/global economic environment (Clark 2001). The seat capacity of an aircraft is rather inflexible from the schedule planning perspective, although aircraft configuration changes can alter the capacity marginally (but not quickly, say in days). Hence, matching the uncertainties of passenger demand with the fixed aircraft capacities of various fleets is always a challenging task for airline fleet assignment.

From historical booking records, airlines can build demand "profiles" on a leg or an origin-destination (OD) basis. A leg basis views the demand of a leg (i.e. a specific flight) as the demand between two airports, while the OD basis adopts a wider view which considers the "direct" demand between an OD pair, as well as other possible forms of "indirect" demand between the same OD pair, via other transit points in a network. Based on demand forecasts and available fleets, an airline schedule is partitioned into a number of sub-networks, corresponding to different fleet types. The collation of flights in all sub-networks comprises the full schedule, while each flight is only assigned to a fleet (a sub-network) without overlapping among sub-networks of different fleets (Bazargan 2004). Fleet assignment optimisation models are usually configured to maximise the profitability of airline schedules by considering stochastic passenger demands and fleet capacity allocation such as the work by Barnhart et al. (1998), Hane et al. (1995), and Jarrah et al. (2000).

4.1.2 Aircraft Routing and Routing Constraints

Following fleet assignment, those flights assigned to a specific fleet type/family are assigned to individual aircraft for operation. This assignment is called *aircraft routing* (also known as aircraft rotation or tail assignment by airlines), because each line of "routing" is identified by the "tail number" of an aircraft (Bazargan 2004). The objective of aircraft routing is to allocate a limited number of aircraft of the same fleet type/family to conduct those flights assigned earlier by fleet assignment. There are three main constraints to aircraft routing: (a) flight continuity, (b) flight coverage, and (c) maintenance requirements.

Each line of routing is designed for operation by an individual aircraft and each line comprises a number of flights that are formed in a chronological order, so that flight continuity ensures that the same aircraft can operate those flights in the specified order. In addition, the following two requirements are needed to satisfy flight continuity: the destination airport of a predecessor flight is the origin airport of a successor flight; and, the time gap (i.e. the turnaround time) between the scheduled arrival time of a predecessor flight and the scheduled departure time of a successor flight is longer than the minimum turnaround time of a specific fleet type. Figure 4.1 illustrates flight continuity in aircraft routing. On Route X, flight SS001 departs at Airport A and arrives at Airport B, where the next flight on the same route, i.e. SS002 departs. The time gap between the arrival time of SS001 at Airport B and the departure time of SS002 is the scheduled turnaround time of SS002 and is represented by a flat line along the time-line of Airport B. This time gap should be greater than or equal to the minimum turnaround time required to turn around the aircraft at Airport B.

The flight coverage constraint in aircraft routing is to ensure that any flight on the timetable is at least (and at most) operated by a fleet type and thus, is "covered" in a route in a sub-network of aircraft routing. Using the example in Figure 4.1 above, eight flights are assigned to this fleet type and form a sub-network which has two routes, i.e. Route X and Route Y. These eight flights are covered by this sub-network and are only in this sub-network.

The third constraint in aircraft routing is the requirement for regular aircraft maintenance. There are four major categories of aircraft maintenance, namely A, B, C and D checks in airline jargon. A checks are carried out the most frequently and the interval between checks depends on the type of aircraft, the maintenance schedules provided by manufacturers, and the regulations of Aviation authorities. In Australia, the Civil Aviation Safety Authority (CASA) is the agency which regulates civil aviation and safety related legislations such as Civil Aviation Regulations (CAR) 1988 and Civil Aviation Safety Regulations (CASR) 1998 (CASA 2008a). Under CAR 1988, an airline (which holds an Air Operator's Certificate) must establish a maintenance schedule that either adopts the manufacturer's maintenance schedule, the CASA maintenance schedule, or an approved in-system of maintenance by an airline.

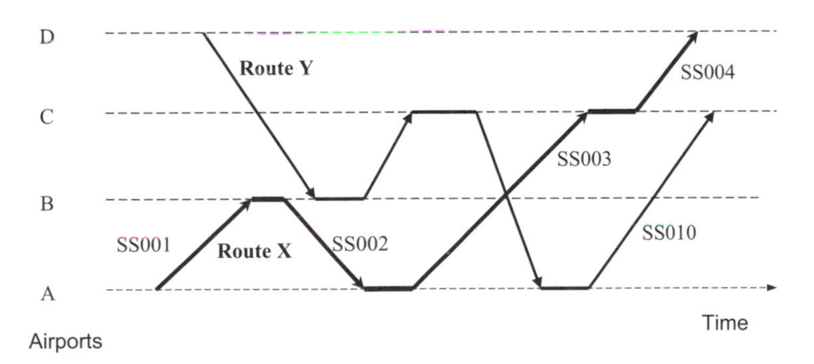

Figure 4.1 Flight continuity and aircraft routing patterns for an example fleet

Aircraft maintenance systems are complex and carefully designed to minimise the risk of mechanical faults of the equipment. In essence, the backbone of a maintenance system is to schedule aircraft for proactive regular inspections, light maintenance, and major overhauls with "labels" such as A, B, C, and D checks (although most modern (post 1990) aircraft only require A and C checks). For aircraft routing, A checks are the most crucial requirement in scheduling, because of the relatively short interval before an aircraft requires an A check. For instance, the modern Airbus A320 family aircraft would need an A check for every 600 flight hours (roughly every six to eight weeks), and a C check for every 6,000 flight hours, or roughly every 18–20 months.

Maintenance schedules may also vary among airlines even for the same type of aircraft, depending on airline preferences and the line maintenance schedule. (Airbus 2008). There is also a "cycle" ceiling (a cycle is a pair of take-off and landing activities) for A checks, together with flight hours and elapsed days since the last check. Whichever cycle, flight hours, or day limit comes first, an A check is required and is often done overnight (6–10 hours) at a maintenance base. Since regular maintenance checks can only be carried out at maintenance bases, an important constraint for aircraft routing is to create overnight "stopovers" for aircraft at maintenance bases, so the required checks can be conducted overnight.

For our example given earlier in Figure 4.1, if the aircraft that operates Route X requires an A check by the end of the planning period, then Route X needs to connect SS002 with SS010 instead of SS003, so as to route this aircraft to Airport C, where the maintenance base is located. For this new routing arrangement, there will be a long and unavoidable ground time for this aircraft between SS002 and SS010 at Airport A, and consequently this will incur some loss of aircraft utilisation. Alternatively, this aircraft can be scheduled to operate Route Y, which ends the routing period at Airport C (just in time for an overnight A check), and then will also have high aircraft utilisation hours during the routing period. The other aircraft can then operate Route X without any changes to routing plans. However, when the fleet size and the sub-network grows, simple solutions like the aforementioned are not easily available and airlines usually rely on aircraft routing models for optimised solutions (Barnhart et al. 1998; Clarke et al. 1997; Cordeau et al. 2001; Pauley et al. 1998).

4.1.3 Aircraft Routing Network – the Backbone of an Airline Network

For airline operations, the network of aircraft routing after fleet assignment is the "backbone" which converts a commercial airline timetable into an operational plan by optimally allocating limited aircraft capacity. Each fleet has a sub-network of aircraft routing such as the one shown earlier in Figure 4.1. The collation of all sub-networks of different fleets forms the aircraft routing network of an airline, which can be significantly large and complex for multiple fleet types and a large number of flights to operate. This is why most large airlines run commercial planning and operation software to assist the planning and operation of large networks such as the Sabre AirOps Suite (Sabre 2008).

Since an aircraft routing network has delicate schedule synchronisation among flights, ground operations, aircraft movements, and fleets, random operational disruptions may damage this well-synchronised network. An aircraft routing network serves as the backbone of airline operations and is further synchronised with the crewing network that jointly facilitates passenger itineraries among airports. Apart from disruptions due to aircraft mechanical problems, disruptions from other sources, e.g. crew, passengers and the environment, can also disrupt aircraft routing and airline operations. In the following sections of this chapter, we will introduce another layer of the airline network, namely the crewing sub-network and further discuss the hidden complexity of an airline network as a whole.

4.2 Crewing and Crew Scheduling

4.2.1 Crew Qualification, Safety Working Hours and Regulations

Crew are required for safe operation of an aircraft for passenger charter. The enroute flight operation of an aircraft is carried out by two groups of crew, namely the flight crew and the cabin crew. The flight crew are technical crew who have received adequate flight training and have accumulated flying hours in order to safely conduct certain flight operations. Two flight crew are required for most modern passenger charter aircraft, including a Captain in command and a First Officer as a co-pilot. Cabin crew are required for providing services aboard an aircraft and more importantly, cabin crew are also responsible for facilitating safety and emergency procedures.

The number of cabin crew needed for a flight depends on the workload required for the level of services provided aboard an aircraft. Often we see that there are less cabin crew for a flight operated by a low-cost carrier (LCC) than one by a network carrier (NC), due to less bundled services provided for passengers on a LCC flight. However, there is a minimum number of required cabin crew aboard an aircraft (charter or regular passenger transport), in terms of cabin crew and passenger ratio for safety reasons. Civil Aviation Orders (CAO) by CASA in Australia set the limit that one cabin attendant is required for every unit of 36 passengers for an aircraft with seat capacity higher than 36 and lower than 216, as seen in section 20.11 and section 20.16.3 of CAO-20 (CASA 2008b).

For safety consideration and fatigue management, the working hours of flight crew are regulated by civil aviation authorities around the world. In Australia, the CAO by CASA also regulates the flight time limitations of flight crew in Part 48, which is also known as CAO-48 in the industry. For instance, a flight crew cannot be rostered in excess of 900 flight hours in 365 consecutive days, or 100 hours in 30 consecutive days, or 30 hours in seven consecutive days. However, CAO-48 does not regulate the flight time limitation of cabin crew. Instead, cabin crew flight time is constrained mostly by Enterprise Bargain Agreements (EBAs) negotiated between cabin crew unions and airlines. Some example EBAs between Australian carriers and crew unions can be obtained online from the Workplace Authority of Australian Government at http://www.workplaceauthority.gov.au. For instance, Australian Airlines' (a former subsidiary of the Qantas Group until 2006) cabin crew EBA enforces a limit that the total duty hours for any cabin crew shall not exceed 1,365 duty hours per annum. For a roster period of 28 days for Australian Airlines, the maximum duty hours was 125 hours, and 43 duty hours was the upper limit of duty hours in any consecutive seven-day period.

EBAs between airlines and crew unions may be based on flying time limits or duty hour (also known as "credit hour") limits. For the example of Australia Airlines, flight crew were rostered and paid on the basis of flying hours, but cabin crew were rostered and paid fully on the basis of credit hours. The "credit hours" of a duty for a cabin crew include sign-on/sign-off time, on-duty rest time, flying time and transit time between flights. Thus, crew rostering becomes a complex exercise due to the calculation of credit hours and the limitations imposed by various EBAs. This complexity also introduces some substantial uncertainties for long-term crew resources planning, which is an essential part of airline business, given that crew cost is usually one of the most expensive items (second only to fuel cost) for an airline.

4.2.2 Crew Base and Crew Scheduling

Apart from flight time limitations, the location of a crew base has important economical and operational implications for crewing and crew scheduling. A *tour of duty* (TOD) for a crew member is defined as: *starting from the sign-on time at the crew base until the sign-off time at the same crew base*. Hence, a TOD may last from one or two days for a domestic crew, to a number of days for an international crew, depending on flight timetables and airline networks. If a TOD involves overnights at ports other than the base of a crew, airlines will incur other crewing expenses such as accommodation, ground transport and meal allowances. Moreover, if crew are paid based on credit hours, then the "length" of a TOD would significantly influence the "cost" of crewing. Therefore, the common objective of crew scheduling is to minimise the total cost of crewing, which includes crew salaries and other crewing expenses.

The location of a crew base depends on an airline network, the shape and size of the network (in terms of airport numbers and locations), crew categories (reflecting fleet types) and cost considerations, e.g. crew salaries at different bases. The common practice of establishing a crew base for an airline is to base most crew at its hub airport (or at the headquarter port of an airline) with other crew based at strategic locations, with operational or economical benefits in a network. We will discuss the operational benefits of a crew base later when we discuss crew scheduling.

Regarding the economical benefits of setting up a crew base, this is often realised from setting up an overseas crew base, where the cost (mostly salary rates) of employing a crew member is significantly lower than the home country of an airline. For instance, the international operation of Qantas Group has a crew base at Sydney (where Qantas is based) and a crew base at Bangkok. Pay rates for locally employed crew at Bangkok are lower than those in Australia, providing significant cost saving for Qantas international operations. In addition, crew based at Bangkok can be assigned to those TODs connecting Southeast Asia and European destinations in the Qantas network, thus providing shorter TODs for crew and a cheaper cost base for crewing. The operation between Southeast Asia and Australia can be both operated by crew in Thailand or local crew in Sydney.

There are two major steps for crew scheduling, namely crew pairing and crew rostering. Crew pairing (or crew pattern generation), similar to aircraft routing, is to create a number of crew pairings that are economically efficient (the lowest cost possible) and comply with workplace EBAs and relevant government regulations. Flight crew can only operate one type of aircraft at a given point of time, due to "type endorsement" regulations. Thus an A320-endorsed pilot cannot operate a B737 aircraft, even though those two types of aircraft are similar in size. Cabin crew, on the other hand, are more flexible and can be assigned to various types of fleets, as long as a cabin crew has received the required safety training for certain aircraft types.

Following the previous aircraft routing example provided in the previous section, a corresponding crew pairing solution is presented in Figure 4.2, in conjunction with the planned aircraft routing plan. Two aircraft of the same fleet type are needed to operate eight flights. Crew pairing results show that there are three crew pairings that comply with working hour limitations and other constraints specified in EBAs. Pairing No.1 connects SS007, SS008 and SS004, in which an aircraft change takes place between SS008 and SS004 (from Route Y to Route X) and a on-duty rest period is also involved. Pairing No. 2 includes SS001, SS002 and SS003. Due to working hour limits, SS004 on Route X is operated by Pairing No. 1. Pairing No. 3 includes SS009 and SS010, which are from the second half of Route Y. We can see from this example pairing solution that at least three pairings are required to "cover" all flights (flight coverage), and the flight "continuity" requirement is also reserved within individual crew pairings, just like in aircraft routing. In addition, there is a close relationship between aircraft routing plans and crew pairing generation, because the construction of crew pairings is based on aircraft routing solutions. The close synchronisation of these two networks will be further discussed in the following sections of this chapter.

Crew pairings, so far, are not yet assigned to individual crew members. This assignment only occurs during crew rostering, which is the process of building up detailed rosters for individual crew within a specific roster period. A TOD for a crew may contain a few pairings over a number of days. A TOD, by definition starts and finishes at the crew base and can not exceed the limit of working hours in seven consecutive days. After a TOD, a required minimum rest time (in hours

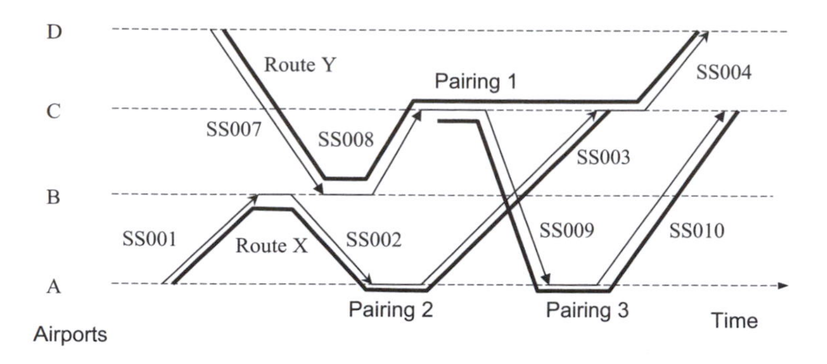

Figure 4.2 Crew pairings in conjunction with aircraft routing

or days) is scheduled to allow crew to recover from duty, especially if a TOD involves long-haul flights that have more than a four-hour time zone difference from the crew base. Each crew receives a detailed roster for a roster period, and this roster is also the basis on which crew salary is calculated (either on flying hours or on credit hours). For most EBAs in the airline industry, a minimum of paid flying hours or paid credit hours is often required, so a minimum wage is guaranteed for any roster periods.

4.2.3 Crewing Network and Synchronisation with Aircraft Routing Network

It is seen in Figure 4.2 earlier that the crewing network for a specific fleet is closely synchronised with the corresponding aircraft routing network. Due to this synchronisation, aircraft routing and flight operations are influenced by the reliability of crewing networks, and vice versa. Delays may occur to the crewing sub-network, due to sickness, late crew connections between flights, or delays due to passenger connections at airports. For instance, if flight SS008 is delayed due to passenger check-in and baggage loading, then SS009 may be delayed because both flights are on the same route, i.e. Route Y, which is operated by the same aircraft. Crew operating for Pairing No. 1 can be late (due to late SS008) for SS004, affecting the operation of Route X by another aircraft.

4.3 Complex Airline Networks

4.3.1 Complex Network Operations

Flights are connected in an airline network (as well as between the networks of different airlines for code-sharing operations) through four types of "flow networks" including: the passenger itinerary network (including passenger baggage transport), aircraft routing network, cargo/goods shipping network, and the crewing network. While an airline designs a network and accordingly the aircraft routing and crewing in the network, passenger itineraries and cargo/goods shipping are influenced and driven by stochastic market demands among the airports in the network.

Earlier in Chapters 2 and 3, we briefly discussed the complexity of airline operations at airports. This complexity mainly comes from the time pressure of conducting aircraft turnaround operations, passenger (and baggage) transfer among flights, and crew connections among flights at an airport. On a network scale, the complexity of airline operations extends from operations at individual airports to the operations of flights between airports in a network. Unlike the operations at airports, operations in a network cannot be fully managed by focusing only on individual airports, because operations at an airport may cause disruptions to operations at other airports in the same network, due to "flows" in the network. This complexity is best illustrated by Figure 4.3.

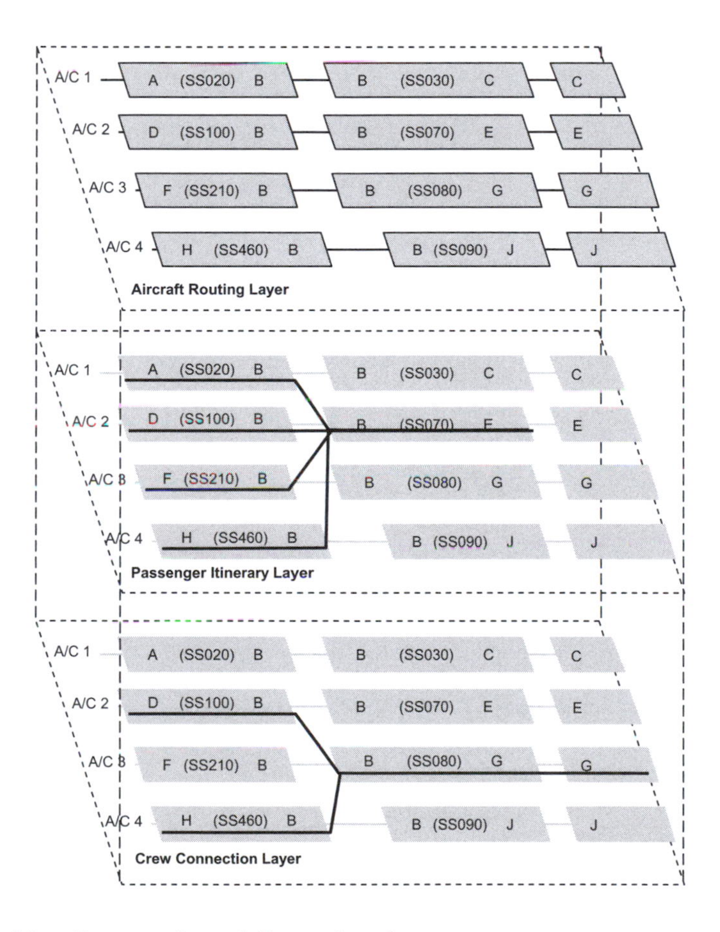

Figure 4.3 Layers of an airline network

On the "layer" of aircraft routing in Figure 4.3, flights on a timetable of an airline are operated and connected by individual aircraft of different types, depending on the result of fleet assignment and aircraft routing. In operations, delays to a flight may cause delays to other flights, if those flights are operated by the same aircraft. For example, aircraft No. 1 (denoted by "A/C 1") operates flight SS020 from airport A to airport B (which is the hub airport in this example), then SS030 from airport B to airport C, and so forth. In the example network illustrated in Figure 4.3, there are four aircraft in the routing layer of the network and there are four flights from spoke airports (A, D, F and H) to the hub airport (B) during the inbound "wave" of the schedule.

On the layer of crew connection, the crew (both flight crew and cabin crew) follow aircraft routing and airline schedule, as well as working under the safety regulations of aviation authorities and the limits of individual airline EBAs. Working conditions for crew are stricter than for aircraft routing for safety and fatigue management reasons. For example, we see from Figure 4.3 that flight crew transfer from flight SS100 to flight SS080, while cabin crew connect from SS460 to SS080 during aircraft turnaround time. Since crew arrive with the aircraft and need to connect to other flights (according to crewing plans) during aircraft turnaround time, it is possible that outbound flights, e.g. SS080 may be delayed due to late connecting crew from SS100 and SS460. The "sit time" required for a crew connection at an airport is usually longer than the minimum turnaround time of an aircraft; however, it is still possible that a flight incurs departure delays due to late inbound crew connections.

The passenger itinerary layer, as shown in Figure 4.3 is the most complex one in an airline network. In a strong hub-and-spoke network, an inbound flight may bring passengers who will connect to a number of outbound flights at a major hub airport. Similarly, an outbound flight, e.g. SS070 as illustrated in the example, may receive connecting passengers from a number of inbound feeding flights such as SS020, SS100, SS210 and SS460. In a typical "wave" of inbound flights in a hubbing schedule, there could be ten to twenty flights, depending on the intensity of hubbing by an airline (Burghouwt 2007). Inbound flights carry passengers with different itineraries to a hub airport where these passengers can then connect to different outbound flights according to their planned travel itineraries.

The higher the hubbing intensity of a schedule at a hub airport (i.e. more inbound/outbound flights in a wave), the more connection opportunities a schedule can create at the hub. Furthermore, the checked-in baggage travelling with passengers also need to "connect" between flights. Depending on how transfer baggage is handled at an airport, the minimum connection time between flights also needs to consider the time required for baggage processing. Hence, it is likely that passengers may be able to transit to an outbound flight even if an inbound flight is delayed, but the passengers' baggage is left behind waiting for the next connecting flight to the same destination. This will surely cause travel inconvenience to passengers and extra operating costs to an airline.

In a similar context, cargo/goods shipped in an airline network "flows" the same way as passenger bags do, but these may have more complex "itineraries" than passengers do. This is because there are usually fewer cargo processing facilities in an airline network (only a few cargo hubs) than passenger transfer terminals. In addition, most cargo shipments do not require specific departure flights or arrival flights, as long as goods arrive at the destination port within a desired delivery time (with the exception of express cargo). Hence, it is often seen in the industry that a cargo shipment may travel through a few flights via a central cargo processing hub before arriving at the destination. Although this complex cargo itinerary may seem to add more complexity to an airline network, it does not cause as many disruptions as passenger transport, due to the nature of cargo shipment and the lower on-time arrival targets (with the exception of express cargo). In other words, ordinary cargo can tolerate more delays due to the flexible delivery target time, and may go through a longer route between the origin and destination airport. Express cargo operates in a similar fashion to passenger charter, due to the tight delivery time window specified by customers. However, express cargo shipments can be transported and processed during the night, when airports are less busy.

Some properties of airline networks are desirable from the commercial and financial perspectives of airline business. First, given the high capital investment in airline assets such as aircraft, the financial pressure to make profits naturally drives airlines to adopt a hub-and-spoke (or similar) network in order to extend the market "coverage" of a network in terms of geography and market segments. Hence, small spoke cities with low demand between each other can be served via larger hubs/cities in a network. This development started in the 1970s when the U.S. aviation market was deregulated, causing the emergence of two distinctive network strategies, namely the hub-and-spoke system by major carriers and the point-to-point system by low-cost carriers (Burghouwt 2007). For network carriers (and indeed for the full-service airline business model), the hub-and-spoke system creates a network in which airlines can vary the "supply levels" easily in the network by using different types of aircraft and altering flight frequency in order to meet uncertain travel demand among cities. This creates the desirable "economies of density" for a hub-and-spoke type of network; the higher the flight density is in the network, the lower the marginal unit cost becomes, hence the higher the profits (Button 2002). This "property" of an airline network is highly desirable from the commercial perspective of the airline business.

Second, a less attended to property of the airline network emerging from the U.S. deregulation is the potential of point-to-point networks. Instead of creating a complex network for connecting traffic, low-cost airlines such Southwest Airlines in the U.S. focus on "trunk routes" where high demand between airports exists. This type of airline network also creates some desirable properties. Among them, the most essential one is the "simplicity" of airline products, airline operations and the airline business model, resulting in low unit costs and high profitability for airlines. In a point-to-point network, the "economies of density" exist on a flight level but not on a network level. Hence, low-cost carriers focus on the high-volume traffic between airports, although this operation may have lower yields than the one by network carriers on the same route. This property of point-to-point networks is highly desirable from the financial perspective and has sparked a few airlines adopting similar network models outside the U.S., such as Ryanair and easyJet in Europe, Virgin Blue in Australia and Air Asia in Southeast Asia.

4.3.2 Operating Problems Emerging from Complexity

Although different types of airline networks bring different desirable properties to airline business, some undesirable properties are unavoidable and may eventually incur higher costs that could outweigh benefits from network "synergy". For a hub-and-spoke network, most undesirable properties stem from the operational characteristics of such a network. First, a hub-and-spoke system causes low utilisation of airline and airport facilities such as airport gates, scarce runway slots at busy airports and airline ground staff. Flights in a hubbing schedule arrive at a hub airport in a "wave" form, creating peak demand for airport and airline resources at an airport. During off-peak hours, airport facilities and airline resources are under utilised and mostly idle. This inefficient utilisation of airline resources and airport facilities increases operating costs of airlines as well as airports.

Second, a hubbing network often causes lower aircraft utilisation than that in a point-to-point network. The "utilisation" of an aircraft is measured by the average flying hours an aircraft can do for each day. Inbound flights at a hub airport often spend a longer time on the ground (longer turnaround time) in order to create flight connection opportunities among each other. As demonstrated in Figure 4.4, we can see that within the "connection window" there are four possible connections for flight No. 1 and a maximum of 16 potential passenger connections among the four inbound flights and four outbound flights at the hub airport. The connection window is often used as a proxy of the maximum time that passengers are willing to spend waiting for flight connections at an airport. The length of this time window also depends on the minimum turnaround time needed for specific aircraft types by an airline. In addition, this time window also depends on the minimum connection time imposed by the airport operator, which by and large depends on the walking distance between gates and terminals, the minimum turnaround time of aircraft, and the time needed for processing connecting baggage.

Individual turnaround time for those four aircraft in our example are longer than the minimum turnaround time needed. In reality, it may not be always possible to have the full 16 connections available on a commercial basis (some city pairs have higher demand than others), so some aircraft can have a shorter turnaround time at the hub, so as to improve aircraft utilisation. It is noted that in the above example, if inbound passengers on flight No. 4 were to connect to outbound flight No. 5, the time gap between the arrival of No. 4 and the departure of No. 5 must exceed the minimum connection time requirement. The trade-off we can observe here is between flight connection opportunities (commercial benefits) and aircraft utilisation (operating costs). However, in a point-to-point network without "on-line" passenger connections, the example in Figure 4.4 will only provide eight individual flights operated by four aircraft, because no on-line flight connections are provided among flights. Hence, aircraft can be turned around in a relatively short time, so as to achieve higher utilisation. This practice is widely adopted by low-cost carriers in order to create the essential cost advantage that low-cost carriers enjoy over network carriers (Gillen and Lall 2004).

Third, the complex nature of airline networks in a hub-and-spoke system may compromise the operating performance during operations, especially when facing unexpected disruptions. The state of a hub-and-spoke system can be disrupted by minor delays due to passenger connections, crew connections and aircraft routing. Network carriers often need to make tactical decisions during operations as to whether an outbound flight waits for connecting passengers or not. Missing connections for passengers cost extra for airline operations and also cause inconvenience for passengers. As for major disruptions such as inclement weather or aircraft technical problems, hub-and-spoke systems can be seriously disrupted and we often see airlines cancelling hundreds of flights due to major disruptions to hub operations. Costs arising from managing a "compromised" network are very high, especially for a hubbing system. This operational concern triggered the recent action of "de-peaking" by Lufthansa, United Airlines, and American Airlines, in order to reduce operating costs from those undesirable properties of hubbing systems (Goedeking and Sala 2003).

For point-to-point networks, some undesirable properties also exist; most of which are associated with airline operations and have cost concerns. Airlines (often low-cost carriers) tend to schedule more flights for aircraft, taking advantage of the operational simplicity of the low-cost business model (Gillen and Lall 2004). This simplicity comes from no or little passenger on-line connections (hence no or little baggage connections) among flights and the "no-frill" business model, so there is less workload for aircraft turnarounds. This simplicity reduces the time required for turning around aircraft and increases the daily utilisation of aircraft.

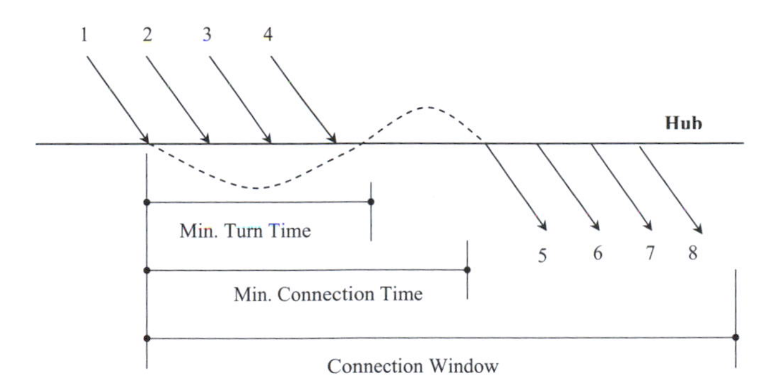

Figure 4.4 Flight connections at a hub airport

Seen as an operational and financial advantage, short turnaround time also brings negative impacts on airline operations such as tight turnaround operations and delay propagation between flights.

Although carrying significantly less connecting passengers than network carriers, aircraft and crew of low-cost carriers still connect on the routing and crewing layers of a point-to-point network. Due to tight aircraft routing plans, it is more likely that flights are delayed due to aircraft routing and crew connections. Delays in such a situation can be more costly than in a hub-and-spoke system, in which aircraft tend to have more turnaround buffer time to absorb unexpected delays. Delay propagation can occur quickly among flights in a point-to-point network due to the short turnaround time between flights, and airlines often need to cancel flights, delay flights or swap aircraft during operations in order to maintain schedule integrity.

Some operational issues one can see in daily airline operations are in fact "created" by airlines because of their pursuit of those desirable properties of airline networks. The weakness of the hub-and-spoke system is not only because of airline operations or unexpected disruptions, but also because of the adoption of such a network. This weakness is inherent with any similar networks which resemble a hub-and-spoke structure as seen in the power industry and telecommunication networks (Barabasi 2003.). Clearly, if an airline adopts a point-to-point network and avoids complex passenger interlining activities, the network will not suffer from similar weakness as a hub-and-spoke network, although it still suffers from the common weaknesses of a "connected" network such as delays. Therefore, when studying airline operations, one must not ignore the big question: *Where do these operational problems come from?* While fixing operating procedures may fix operational issues in a short-term, similar or new operational issues may emerge quickly in a network, if the fundamental question is not addressed and investigated fully.

4.3.3 Measuring Airline Network Complexity – A Schedule Planning Perspective

4.3.3.1 Spatial Attributes Airline network studies often use the Gini index as the primary indicator of the spatial concentration of an airline network (Burghouwt 2007; Reynolds-Feighan 1998), although a number of other indices have also been used in the literature, e.g. the Herfindahl index, the valued-graph index by Shaw (1993), the beta index of spatial concentration by Chou (1993a; 1993b), and the gross vertex connectivity by Ivy (1993). In the context of aviation, the Gini index is often defined to evaluate the spatial concentration of an airline network (in terms of traffic) as in (4.1):

$$G = \frac{1}{2n^2\overline{y}} \sum_i \sum_j \left| y_i - y_j \right| \tag{4.1}$$

where y_i/y_j represents the traffic of airport *i* and *j* of the study airline, in terms of weekly seat capacity; n is the number of airports in the network of the study airline. Thus, the Gini index has the minimum value of zero when all the traffic of an airline is evenly distributed across the network, and the Gini index has the maximum value of $G' = 1 - \frac{2}{n}$ (where $n > 2$).

Due to the differences on the scale of the Gini index when comparing networks with different sizes, Burghouwt (2007) proposed a normalised Gini index, called the Network Concentration (NC) as in (4.2) below. Hence, NC has a value from zero to one; a value of one represents a network with most traffic concentrated on a "trunk" route, while values less than one indicate the spreading of traffic within a network. A major defect of NC as pointed by Burghouwt (2007) and Martin and Voltes-Dorta (2008) is that the same NC value may correspond to different network configurations, i.e. NC values and network configurations do not have a one-to-one unique mapping relationship.

$$NC = \frac{G}{G'} \tag{4.2}$$

4.3.3.2 Temporal Attributes Apart from spatial attributes, airline networks also exhibit temporal attributes and are often classified as "time-varying networks", meaning that the "configuration" of an airline network changes with time. This temporal attribute is due to the fact that there are no "physical" links in an airline network, and "links" only exist when "links" (flights) are scheduled to function at certain times, e.g. scheduled flights in an airline network. Regarding the temporal attributes of a network, one would be concerned how a network is configured to change with time and how this time-varying configuration may influence the performance of such a network during operations. In particular, temporal effects are more significant and obvious in a hubbing schedule, which contains alternating inbound and outbound "waves" of flights. This wave structure and its temporal attributes define most hub-and-spoke airline networks in the world as well as the hubbing intensity a network provides.

The main purpose of adopting a hub-and-spoke network (regardless of the strength of hubbing) is to create flight connection opportunities among spoke airports via hub airports. Hence, the temporal attributes of an airline network also reflect the connections provided via hubs in a network, hence determining what tickets among markets an airline can offer. The total travel time between origin and destination airports (including transit time among flights) and the routing distance of the whole journey are two key factors in determining the commercial success (saleability) of flight connections, although other factors also play a role such as air fares, airport facilities, departure/arrival time, frequent flyer membership, and passenger preferences in travel. To capture the temporal attributes of airline product offerings in a network, a number of measures have been developed in the past. Burghouwt (2007) developed the *Weighted Indirect Connection* (WI) index that reflects both the time and distance attributes of a journey between two airports. WI was defined as follows:

$$WI = \frac{2.4 * TI + 1.0 * RI}{3.4} \tag{4.3}$$

where $TI = 1 - {}^{T_h}\!/_{T_j}$ represents the "transit index", and T_h / T_j denotes the transit time of a journey at the hub and the maximum connection time of the journey respectively; $RI = 1 - (2.5 * R - 2.5)$ represents the "routing index", with R denoting the ratio between the journey time with connections and with direct flights. The weight factors in (4.3) were based on the assumption that the value of time is 2.4 times higher for transit time than for block time (i.e. the actual time a person spends aboard) (Burghouwt 2007). Adding up all *WI* indices of potential connections offered in a network, one would generate the overall measure of indirect connections, namely *WNX*.

Hsu and Shih (2008) developed the *Mobility Index* and *Accessibility Index* based on the emerging theory of networks, especially the "small-world network" concept by Watts and Strogatz (1998). The Mobility Index was developed to measure the temporal attribute of an airline network, and the Accessibility Index was to reflect both the temporal attribute and the spatial attribute (i.e. market concentration or attraction) of an airline network. The Mobility (M_G) and Accessibility Indices (A_G) were modelled respectively by (4.4) and (4.5) (Hsu and Shih 2008):

$$M_G = \frac{1}{N(N-1)} \sum_{i \neq j \in G} \left(\frac{1}{t_{ij}} \right) \tag{4.4}$$

where N is the number of nodes (airports) in network G; t_{ij} denotes the shortest travel time between airport pair *i-j*; ${}^{1}\!/_{t_{ij}}$ represents the mobility.

$$A_G = \sum_{i \in G} A_i = \sum_{i \in G} \sum_{i \neq j \in G} \left(\frac{P_j}{(t_{ij})^{\alpha}} \right) \tag{4.5}$$

where A_i denotes the accessibility of airport *i* in network G; P_j is the airport traffic of *j*; t_{ij} is the travel time and α is a decay parameter. Both sets of measures from Burghouwt's and Hsu and Shih's work for different purposes. As demonstrated, the *WNX* index could be used to evaluate the time-varying connectivity of an airline network (Burghouwt 2007, page 72), while the Mobility and Accessibility Indices were shown effective on assessing network efficiency, as shown in the case study of airline alliance networks by Hsu and Shih (2008).

4.3.3.3 Pitfalls of these Approaches Given the above network measurement indices, there are still pitfalls in using those indices. In a recent study by Martin and Voltes-Dorta (2008), they showed theoretical evidence of why some common measures of airline network complexity, e.g. the Gini index were not suitable in measuring airline networks and that the use of these indices may produce misleading results. The emerging key point from the work by Martin and Voltes-Dorta was that the definition of a "hubbing" network/schedule, regardless of its intensity should be based on passenger connections in a network, rather than simply on the airline schedule structure or leg-based traffic data. As pointed out earlier by Button (2002) and debated by Martin and Voltes-Dorta (2008) more recently, the lack of a definition of a "hub" airport has caused some confusion and misunderstanding in airline network studies. Recently, the more accepted definition of a hub airport is based on: the passenger traffic connection percentage among flights offered by airlines at a study airport. This definition excludes the possibility that a pure point-to-point network with high concentration of flights and traffic at an airport is mis-classified as having a "hubbing" schedule at this airport, due to the fact that no passengers connect between flights in a pure point-to-point network.

To further clarify the definition of hubbing, readers must have a clear view of passenger traffic measurements both on a "leg-based" and on a "journey-based" (also called OD based) level. To illustrate the difference of leg-based traffic and OD (origin-destination) based traffic, we use Figure 4.5 as an example. In Figure 4.5, there are two airline networks which operate among four airports with similar schedules. Network I operates a pure point-to-point schedule, in which the traffic of A-B, B-C and B-D are 150 passengers each, if measured on a leg basis. Network II operates a hubbing schedule, in which passengers from A can fly to B, C and D via the hub airport, B. On a leg basis, the traffic volume of A-B, B-C and B-D are 150 passengers for Network II as well. However, if the traffic is calculated based on passenger journeys between ODs, figures stay the same for Network I (ignoring any passenger "self-hubbing" traffic), but figures are quite different for Network II due to connecting passenger traffic via B. OD traffic of A-B, A-C, A-D, B-C and B-D are 50, 50, 50, 100, and 100 respectively.

The implication of passenger connections for airline resources allocation (in particular for fleet assignment) is that a single fleet (150 seaters) can operate both networks without offering direct flights between A-D and A-C. If direct flights are needed for A-D and A-C routes, different fleet types (one 50 seater and one 100 seater) will be needed for Network II, because the traffic between A-B reduces from 150 to 50 and the traffic for B-D and B-C routes are now only 100 passengers, insufficient for an aircraft with 150 seats. This example also demonstrates why a "hub-and-spoke" schedule in Network II can generate economies of density, a natural outcome of network economies and efficiency.

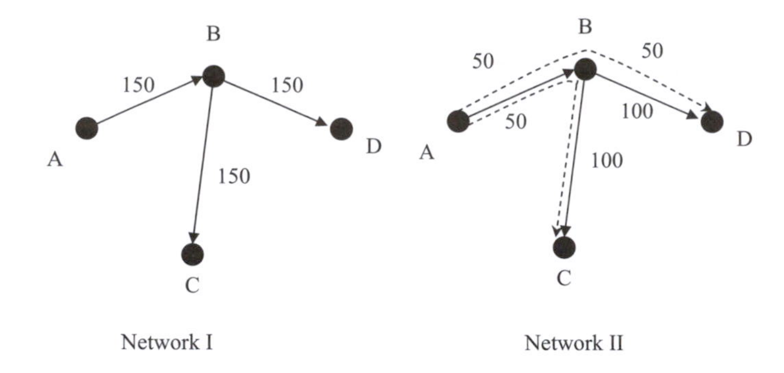

Figure 4.5 Different passenger flow patterns on the same network structure (adapted from Martin and Voltes-Dorta (2008))

Given the pitfalls of existing indices, Martin and Voltes-Dorta (2008) developed the *Hubbing-Concentration Index* (HCI), combining measurements of connecting passenger shares at a hub and the routing concentration of passenger journeys in a network. HCI was modelled by (4.6) below:

$$HCI = \sum_{i \in G} \sum_{j \in G} \left(w_{ij} H_{ij} \right) \tag{4.6}$$

where $w_{ij} = {}^{p_{ij}}\!/_{\mathrm{P}}$ represents the weight of traffic between OD pair *i-j*, a ratio between passenger traffic of *i-j* (p_{ij}) and total traffic of the airline in the network (P); $H_{ij} = c_{ij} \sum_{i \in G} s_i^2$ models the passenger connecting share of *i-j* market (denoted by c_{ij}) weighted by Herfindahl Index of the market share of the airline at airport *i* (denoted by $\sum_{i \in G} s_i^2$). The HCI index prevents the pitfalls of other network measurement indicators and was reported to provide superior information in airline network studies as seen in the case study given in Martin and Voltes-Dorta (2008).

A difficulty of using the HCI index in studying airline networks is the availability of passenger connection data. In the U.S., this data is available from the "Origin and Destination Survey" database of the Bureau of Transportation and Statistics, which samples 10 per cent of all ticket sale records in the U.S. Beyond this database, passenger connection information is always considered to be commercially confidential by airlines and is not available in other regions except the U.S. This is also the main reason why other network indicators have been more popular in the industry and better developed in the past.

4.3.4 Measuring Airline Network Complexity – An Operational Perspective

Most spatial and temporal airline network indicators are aimed at measuring the geographical characteristics of a network and/or network configurations in terms of flight frequency, flight times and schedule wave structure, but are not designed to assess the complexity of flight connections during schedule execution and how this complexity can be avoided in schedule planning. As suggested by Martin and Voltes-Dorta (2008), the major pitfall of current spatial and temporal network indices in the literature was the lack of recognition (also the lack of data) of passenger connection flows among flights in an airline network. In the context of airline operations, what is essential is to recognise the complex interaction between passenger connection flows, crew connections, aircraft routing and the timetable, so as to develop appropriate measuring indices and evaluate the impact and reliability of individual "links" on airline schedule integrity during operations. Those indicators mentioned in the last section were not suitable for use in evaluating the operational complexity of airline networks due to the fact that those indices were developed to evaluate airline network characteristics initially, and as such are more meaningful for network development and planning purposes.

In light of this, some ad hoc metrics have been developed to fill this gap and these have served as evaluation tools for assessing the operational integrity of individual flights as well as the network as a whole. In a recent paper by AhmadBeygi et al. (2008a), a number of metrics were developed to evaluate the potential impact of delay propagation in an airline network. These metrics included: propagated delay, magnitude, severity, depth (of propagation trees), depth ratio and other ad hoc metrics for assessing crew and aircraft routing plans. AhmadBeygi et al. (2008a) then applied those metrics to two sets of airline schedules by introducing various levels of root delays to individual flights via Monte-Carlo simulations, and investigated how delays may propagate in two different types of networks. This sort of application is often seen in network sensitivity studies, in which one wishes to "test" potential network responses and sensitivities to various "stimulants" (e.g. delays) in different scenarios. The work by AhmadBeygi et al. (2008a) provided a good framework for developing metrics that can be used to study network complexity and sensitivity during schedule planning.

As discussed in earlier chapters, airline schedules are developed to maximise potential profits, so commercial considerations play an essential role in shaping airline schedules, besides resource utilisation such as crew and aircraft. Depending on the structure of airline networks, some flights are planned to connect with others, so as to increase the "coverage" of destinations in a network, as well as to utilise crew and aircraft time. However, in actual airline operations, some connections may be broken due to delays caused by various reasons such as baggage handling delays, connecting passengers, and connecting crew. Based on a similar rationale to the one given by AhmadBeygi et al. (2008a), a set of metrics is developed here that provides airlines and analysts with a tool set to evaluate the scale and impact of disrupted flight connections in a network on the day of operation. These metrics are discussed and modelled in the following sections. Notations and symbols used in this section are explained in the text as well as being listed in the Appendix of this chapter.

4.3.4.1 Scheduled Flight Connectivity To assist metrics development, we will use an example flight, f_{ij} in Figure 4.6 to demonstrate the flight connections in a network. f_{ij} denotes flight i on route j, originating at Airport A and flying to Airport B. There are some potential inbound connecting flights (totalling M_i flights) at Airport A with scheduled arrival times falling within the inbound "connection window" (denoted by CW). Upon arriving at the destination Airport B, f_{ij} may feed connections (such as passengers, crew, and aircraft) to some outbound flights (totalling N_i flights) which have their scheduled departure times falling within the outbound CW at Airport B as illustrated in Figure 4.6.

The size of CW depends on passenger perceptions of transit time as well as the availability of flight connections; the longer CW is, the less saleable a flight connection becomes. To simplify analysis, we will assume a two-hour CW in our example here. Airports (sometimes also airlines) may impose a "minimum connection time" (denoted by MCT) for on-line connections between flights due to ground handling capacity. Flight times with a gap less than MCT is not allowed and is often not available for passenger booking. Inbound flight arrival time, CW, MCT and the scheduled departure time of f_{ij} (noted by s_{ij}^{D}) determines M_i, namely the "inbound connectivity". Similarly, the outbound flight departure time, CW, MCT and the scheduled arrival time of f_{ij} (noted by s_{ij}^{A}) jointly decides N_i, namely the "outbound connectivity". Thus, we define the *Scheduled Flight Connectivity* metric of flight f_{ij} as $SFC_{ij}(M_i, N_i)$.

4.3.4.2 Realised Connection Disruption Although connection opportunities are planned to maximise the utilisation of network economies and efficiency, the realised connecting traffic among flights in a network is stochastic depending on passenger demands and travel itineraries between cities on the day of operation. Among those realised connections, late inbound flights (with connecting traffic)

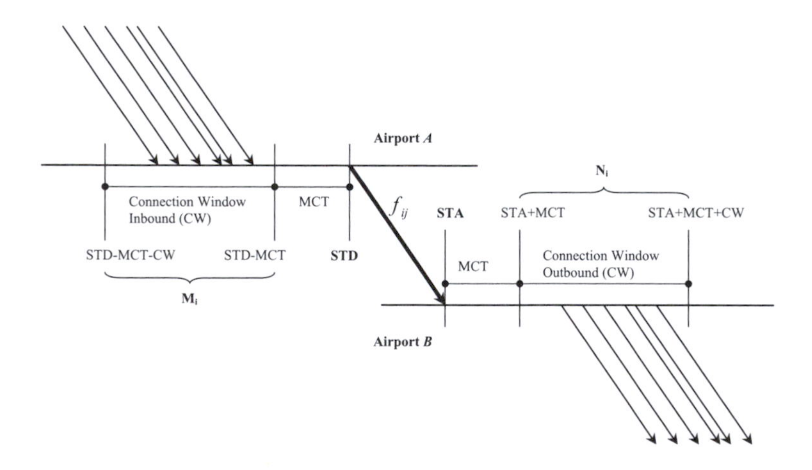

Figure 4.6 Flight connections and connection windows

arriving at Airport A may delay outbound flights to which connecting passengers, crew and aircraft are planned to connect. To evaluate the scale of this realised connection disruption in daily operations, we further define the Realised Connection Disruption (RCD) metric of f_{ij} as $RCD_{ij}(m_i, n_i)$, where m_i is the number of connecting flights inbound Airport A that caused connection delays to f_{ij} (with $m_i \leq M_i$), and n_i is the number of connecting flights outbound from Airport B that received delayed connections from f_{ij} (and $n_i \leq N_i$).

4.3.4.3 Magnitude of Disrupted Connections Delays that propagate between flights due to disrupted connections are also measured to assess the impact of network connectivity. Delays (noted by d_i^m and d_i^n) due to $RCD_{ij}(m_i, n_i)$ are calculated to reflect the magnitude of delays propagated between flights. Hence, we are able to compare the scale of connection disruptions in operations with the planned connectivity of individual flights in a network (SFC); a higher value of RCD means that the planned connectivity was more often disrupted during schedule execution, and logically this has implications for robust airline schedule planning, the emerging scheduling concept we will discuss in Chapters 5 and 6 of this book.

4.3.4.4 Delays from Airline Ground Operations Airline schedules always include buffer time to ensure that minor delays from operational uncertainties can be compensated for without resorting to costly disruption management tactics such as flight cancellation, passenger re-accommodation or aircraft swap. The relationship of delay propagation between flights can be modelled by (4.7). Ground buffer time (${}_{G}S_{ij}^{b}$) and enroute buffer time (${}_{A}S_{ij}^{b}$) are planned to absorb delays from up-stream connecting flights (d_i^m) and ground operational delays (d_{ij}^{OP}).

$$d_i^n = \max\left\{0, \left(\max\left\{0, \left(d_i^m + d_{ij}^{op} - {}_{G}S_{ij}^{b}\right)\right\} + d_{ij}^{ER} - {}_{A}S_{ij}^{b}\right)\right\} \quad (4.7)$$

Operational delays due to airline ground operations (including aircraft turnarounds) may contribute further to departure delays and even arrival delays at the destination airport, if not absorbed by the built-in buffer time. While arrival/departure delays of individual flights can reflect the magnitude of disruptions with respect to planned schedules, delays due to ground operations can further reveal how well a schedule is buffered against stochastic disruptions, and how effective this planned buffer space is in actual operations.

4.3.4.5 Route-based Delays If delays for individual flights, e.g. d_i^m and d_i^n, are summed up on a route basis, this metric, namely the Route-based Delays, reflects the magnitude of delays that are caused by disrupted flight connections on a specific route. d_j^m is the sum of all inbound propagated delays due to disrupted connections (i.e. $d_j^m = \sum_i d_i^m$), and by the same token, d_j^n is the sum of all propagated outbound delays for route j ($d_j^n = \sum_i d_i^n$). This metric is an aggregate metric and can serve as a benchmarking metric when comparison among routes in a network is required. Ideally, one would like to see that routes are subject to similar levels of disruptions in the network. This metric also reflects whether one route is more prone to disruptions than another.

4.3.4.6 Network-based Delays This metric is a further aggregation of those Route-based Delay metrics in a network; d^m denotes the sum of all route-based inbound delay metrics in a network, i.e. $d^m = \sum_j d_j^m$. Similarly, d^n represents the sum of all route-based outbound delay metrics, i.e. $d^n = \sum_j d_j^n$. In particular, this network level metric is useful when one would like to investigate the network status on a daily basis and observe potential seasonal trends.

On the network level, if operational delays (d_{ij}^{OP}) are summed up for all flights in a network as in (4.8), the total operational delays (d^{op}) of the network reflect the magnitude of disruptions that a network is exposed to. This magnitude is influenced mostly by uncertain disrupting events, passenger itineraries, airline operating procedures, and airline resources connections, i.e. crewing and aircraft routing. In conjunction with built-in schedule buffer time, the magnitude of d^{op} should be mitigated due to the planned buffer times in a network. Thus, based on total network delays (i.e. d^m and d^n) and network disruption delays (d^{op}), one would be able to evaluate how well a network is designed (and buffered) against uncertainties in actual operations, and how well the built-in buffer space works when faced with uncertain disruptions.

$$d^{OP} = \sum_j \sum_i d_{ij}^{OP} \quad \forall i \in F \tag{4.8}$$

4.4 Case Study- Measuring Airline Network Complexity

4.4.1 Airline Data

Schedule data and delay data of an anonymous airline (Airline S) in 2004 and 2005 were used for this case study. Delay data (including delay codes) were used to trace the delay trees and delay propagation among flights due to passenger connections, goods loading, crewing, and aircraft routing. Two types of network analyses were conducted: route-based analysis and network-based analysis. Route-based analysis studied the on-time performance of individual routes in the study network, delay propagations between flights, and how delays were absorbed by the planned schedule and buffer time. To minimise the bias due to seasonality, route-based analyses were done for the Northern Summer (April to October) seasons in 2004 and 2005, as well as the Northern Winter (November 2004 to March 2005). Network-based analysis, on the other hand, provided a "longitudinal" view for the status of the study network on a daily basis for both 2004 and 2005. Apart from the easily observed seasonal influence on operations, we will also explore the influence of disruption delays on daily airline schedule execution, and where possible, discuss how well the schedule was designed to withstand stochastic disruptions by using the embedded buffer times in the network.

4.4.2 Route-Based Analyses

Aircraft maintenance routing planning provides optimal routes to cover all flights in the timetable by using limited resources (i.e. aircraft), and meanwhile satisfying aircraft maintenance requirements. Planned routes may be flown as planned during operation, or they may be disrupted and broken due to tactical operating measures, e.g. aircraft swap. Hence, not all routes were flown exactly as planned, and some may be flown more often than others. Schedule and operational data from the anonymous carrier in 2004 and 2005 were used to re-construct operated routes. Since the data were post-operation data, retrieved routes from the data set contained "flown routes" as well as "planned routes". Routes in the Northern Summer flown more than ten times were extracted from the raw data. For each sampled route, say route j, the propagated route delays (d_j^m) of all flights on that route were calculated and statistics of the collected samples were also calculated. Routes in the Northern Winter (from November 2004 to March 2005) were processed for comparison purposes. Hence, two complete Northern Summer seasons (2004 and 2005) and one complete Northern Winter season (2004–05) were available for route-based network analysis.

Results from the Northern Summer operation in 2004 are shown in Figure 4.7 and results of the same season in 2005 are given in Figure 4.8 below. It is seen that the mean propagated route delays in 2004 were slightly lower than in 2005. The mean propagated route delays in the network were about 20 minutes in 2004 and 30 minutes in 2005 but the overall difference was not strongly significant. However, some routes seemed to have higher average route delays, meaning that these routes had more exposure to delays and delays were also prone to propagate along flights on the those routes. Statistically speaking, high mean delays may be the result of extreme samples, and this can be identified by studying standard deviations of the samples, illustrated in Figure 4.7 and 4.8. From the perspective of operations, what Airline S should be concerned about is those routes identified with high mean delay values but low standard deviations, meaning consistent delays in operations.

When the 2004 Northern Winter results in Figure 4.9 were compared with Northern Summer results, it was found that propagated route delays in the 2004 Northern Winter were lower than that of Summer seasons in both 2004 and 2005. The low standard deviation figures on Figure 4.9 showed that operating results (in terms of delays) were more consistent in the 2004–05 Winter season. A possible explanation is that the study carrier (Airline S) operates in the southern hemisphere where the southern summer corresponds with the northern winter in the northern hemisphere. Thus, the rather stable weather condition in the southern summer season may have contributed to the consistent operating results with a lower level of delays across routes in the network.

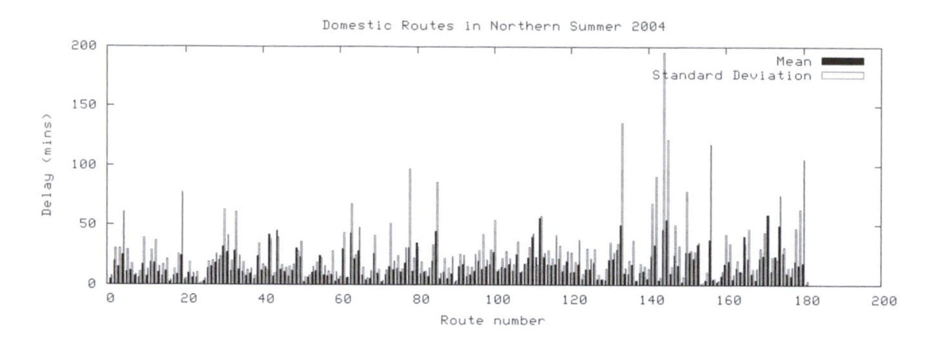

Figure 4.7 Route analyses for northern summer 2004

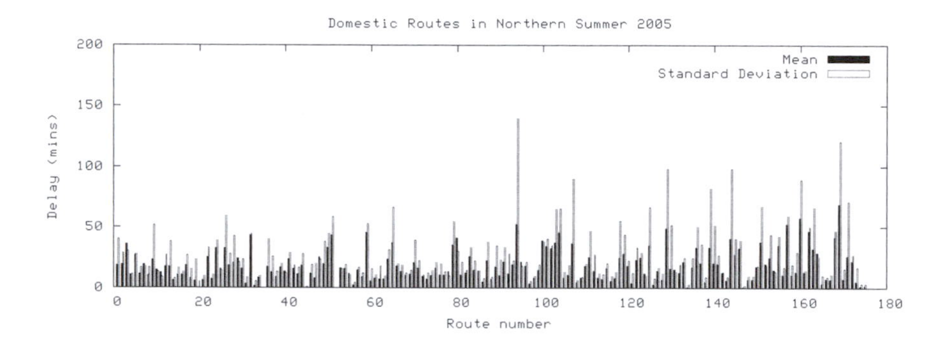

Figure 4.8 Route analyses for northern summer 2005

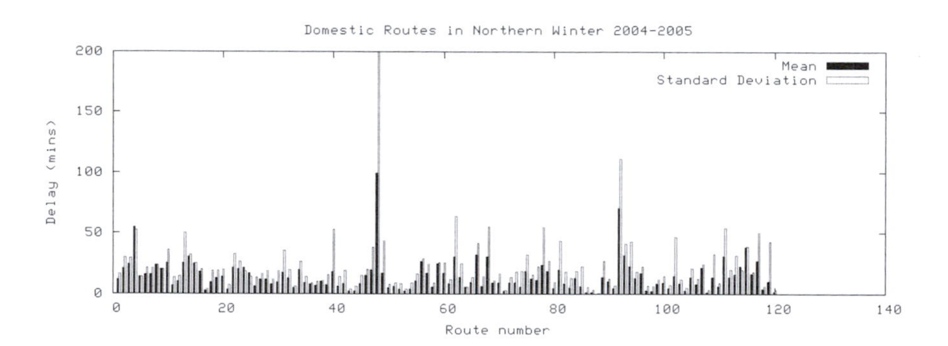

Figure 4.9 Route analyses for northern winter 2004–2005

4.4.3 Network-Based Analyses

The same set of airline data was used to calculate the total daily inbound delays (d^m), the total daily outbound delays (d^n), and the total daily operational delays (d^{OP}) of all flights in the network. The network-wide inbound delays reveal how delays propagate among flights via aircraft routing in the network and whether daily and seasonal effects can be seen from the airline schedule operations. The network-wide outbound delays can reveal how well the planned buffer time has helped mitigate delay propagation among flights due to stochastic operational delays from aircraft ground operations. Comparisons between the network status of Airline S in 2004 and 2005 were conducted and results are shown in Figure 4.10 and Figure 4.11 below.

It can been seen in Figure 4.10 that delay propagation in the network followed a pattern that the total daily inbound delays (d^m, the solid line) of the network were close to but higher than the total daily outbound delays (d^n, the dotted line), due to the embedded ground and enroute buffer time in the schedule. We can also see that the total daily operational delays (d^{OP}, the broken line) of the network were high. These operational delays had been partially compensated for by the buffer time planned in the schedule, resulting in narrow gaps between d^m and d^n as we saw from Figure 4.10.

If we compare the network status in 2004 with that in 2005, we can see from Figure 4.11 that the network status in 2005 was more volatile than that in 2004, in terms of operational delays. It is seen that the propagated delay level in 2005 was generally higher than that in 2004, and this can be attributed to the high level of delays due to disruptions (the broken line in Figure 4.11). This in turn resulted in high departure delays and high propagated arrival delays to other flights across the network.

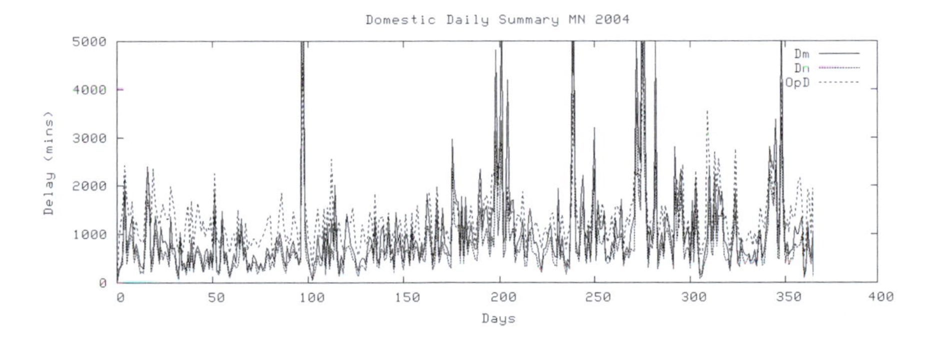

Figure 4.10 Daily network status in 2004

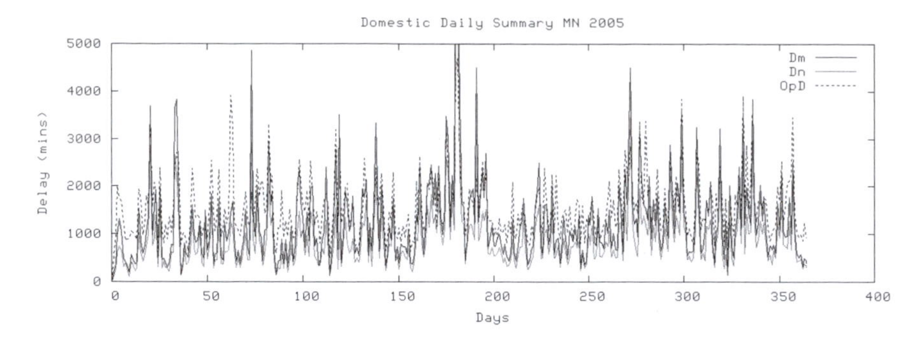

Figure 4.11 Daily network status in 2005

To quantify the differences between operations in 2004 and 2005, basic statistics generating from Figure 4.10 and Figure 4.11 are summarised in Table 4.1. Statistics in Table 4.1 show that the average delays in 2005 were indeed higher than that in 2004. The sharp 25 per cent increase of operational delays in 2005 caused an 18 per cent increase in propagated network delays, and the designed buffer time in the schedule had effectively controlled the delay propagation level in the network; down-stream outbound propagated delays (d^n) were not any worse (although not any better) than up-stream inbound propagated delays (d^m).

Correlation analyses among propagated inbound/outbound delays and operational delays were conducted to test the relationships among these three metrics, and results are shown in Table 4.2. It appears that propagated inbound delays had a very strong correlation with outbound delays as the correlation coefficient was as high as 0.98. Meanwhile, the operational delays also showed a strong correlation with both the propagated inbound and outbound delays (0.80 correlation), revealing strong causal effects among these three metrics. This finding confirmed the common observation in the industry that a well designed schedule should be able to withstand stochastic ground disruptions and delays to a satisfactory level by controlling delay propagation levels in a network. However, the definition of a "satisfactory level" could reveal another gap that exists between the expected level from airline schedule planning and the actual level after schedule execution (Wu 2005). We will discuss the concept of "inherent delays" in more detail in Chapter 5.

Delays in schedule operations mainly come from three main sources: the airspace/airport system capacity constraints, disruptions to airline operations on the ground, and weather conditions. The statistics of delays (sorted by delay code groups) of the 2004 data of Airline S are given below in Table 4.3. We can see that delays due to airline operation related causes represented more than 80 per cent of all delay causes. Some delays were due to operational errors such as aircraft ramp handling, while others were more strongly related to the nature of the airline

Table 4.1 The statistics of daily network status in 2004 and 2005

	Mean Values (μ, mins)		Standard Deviation (σ)	
	2004	2005	2004	2005
d^m	1069	1265 (+18%)	1118	963
d^n	758	894 (+18%)	771	691
d^{OP}	1242	1558 (+25%)	564	664

Table 4.2 Correlation analyses between propagated delays and operational delays

	Correlation Coefficient	d^m	d^n	d^{OP}
2004	d^m	–	0.98	0.80
	d^n	0.98	–	0.78
2005	d^m	–	0.97	0.82
	d^n	0.97	–	0.81

Table 4.3 Statistics of delays sorted by delay code groups

Delay Code Groups	(μ,σ)*	Delay Probability
Others	(36, 30)	1.4%
Passenger and Baggage	(6, 5)	8.9%
Cargo and Mail	(7, 4)	0.6%
Aircraft and Ramp Handling	(6, 5)	11.6%
Technical and Aircraft Equipment	(28, 30)	3.8%
Damage To Aircraft	(26, 23)	0.5%
Automated Equipment Failure	(9, 7)	0.8%
Flight Ops and Crewing	(6, 6)	5.9%
Air Traffic Flow Management	(6, 6)	6.1%
Airport and Government Authorities	(3, 3)	7.7%
Miscellaneous	(10, 8)	0.8%
Reactionary	(12, 11)	6.5%
Inbound Delay	(15, 22)	43.4%
Weather		1.1%

Note: *μ denotes mean values; σ denotes standard deviations.

network design such as crewing and reactionary delays (including aircraft inbound delays).

Statistics from European carriers showed that airline-related disruptions accounted for nearly 50 per cent of all operational delays in 2005 (EuroControl 2005). Hence, a strategy to control network delays from an airline's perspective is to effectively reduce unexpected operational delays in airline operations. This is achievable tactically by improving aircraft ground operation procedures and ground operation co-ordination among ground handling units. Strategically, this can be achieved by improving airline schedule planning to enhance schedule reliability in operations. We shall discuss these schedule improvement tactics and strategies in further details in Chapter 5, when we study disruption management by airlines and schedule optimisation on a network scale.

4.5 Delay Propagation Modelling – Simulation Models

4.5.1 Delay Propagation in an Airline Network

Earlier in Section 4.3 of this chapter, we discussed in detail the complex nature of airline networks due to resource deployment (aircraft and crew), resources utilisation pressure as well as uncertain demand driven air passenger itineraries. Complex networks can be truly complex but also extremely well co-ordinated in operations by design in nature, e.g. an ant's colony, or by thorough planning, e.g. airline networks, air traffic flows or even the global financial networks. Complex networks often do not exhibit the true degree of complexity until the network is executed in an environment in which stochastic disruptions may occur during operations. This is exactly the case for airline networks.

Given the fact that the environment in which airlines operate is subject to stochastic disruptions, the biggest challenge for day-to-day airline operations is to mitigate any negative impacts brought about by disrupting events and maintain schedule synchronisation and integrity. Delays may occur due to disruptions or simply due to lengthy airline operations. Regarding the causes of delays, delays are either classified as "root delays" or "reactionary delays". Root delays are those delays which are directly caused by independent disruptions (thus, they are also called "independent delays" by airlines), and not caused by other delays. For instance, delays due to late check-in passengers are root delays, but delays due to the late loading of passenger bags because of late check-in passengers are not root delays, but "reactionary" delays due to late check-in. Reactionary delays may be caused by the airline's own flight operations, other flights with connecting loads, ground handling operations at an airport, or other up-stream flights at other airports in the same network. This type of reactionary delays on a network scale is called *delay propagation*. By the IATA delay coding system, this source of delays (termed as "reactionary delays" by the system) is represented by delay code numbers 91 to 96.

Delays may propagate in a network via passenger connections, crewing and aircraft routing, unless mitigated by the built-in schedule buffer time or operational interventions by airlines. This delay propagation phenomenon can be significant in a highly connected and synchronised airline network, e.g. a hub-and-spoke network, due to the complex resource and passenger connection at multiple airports. AhmadBeygi et al. (2008a) demonstrated the formation of potential delay propagation "trees" by using data from two U.S. carriers, in light of disruptions, crew connections and aircraft routing schedules. In that study, AhmadBeygi et al. showed how delays can propagate in airline networks and meanwhile, can be potentially controlled by built-in schedule slacks (i.e. buffer time). Although passenger connections were not included in that study, the potential magnitude of delay propagation due to crewing and aircraft routing can still be seen. If passenger connection data were included in delay propagation analysis, we would expect to see more escalation of delay propagation in airline networks, especially for a hub-and-spoke network with more connecting passengers. The scale of delay propagation also reflected how the two airline schedules were planned in the first place.

In daily airline operations, it is essential to maintain aircraft routing assignments due to commercial benefits, operating benefits (crewing matches aircraft types), aircraft maintenance requirements, and controlling delay propagation along routes. To demonstrate delay propagation along routes, a set of delay data from an anonymous European carrier is used in this section. This European carrier (called Airline P hereafter) operated a median network covering Western Europe. The schedule of Airline P was designed for high aircraft utilisation with a single type of aircraft. Turnaround time was typically low, between 20 to 40 minutes for European domestic operations, with a long turnaround time (60 to 80 minutes) for most midday flights at the base airport of Airline P. Clearly, the long turnaround time in the middle of the day was planned as a fire-break time for controlling propagated delays that accumulated in morning operations. In addition, this time window provided Airline P with some aircraft "hot swap" options in midday, if delay propagation went out of control due to disruptions. A route (operated by aircraft No. 8) was chosen for this case study and other follow-up case studies in Section 5.5 of Chapter 5. The average departure delays and arrival delays of this route are shown in Figure 4.12 and flight information is given in Table 4.4.

As seen from Figure 4.12 flights on the study route had departure delays ranging from seven minutes to eighteen minutes (flight PP85) and arrival delays from seven minutes to fourteen minutes (PP85 as well). The close trend between departure delays and arrival delays can be clearly seen, meaning that departure delays had been slightly absorbed by the buffer time planned in the block time of flights. Due to the tight turnaround plans produced by Airline P, inbound delays propagated to departure flights, together with any extra delays due to ground disruptions. This can be seen from the arrival delay of the first flight (PP81) and

Table 4.4 Routing schedule of Aircraft No. 8

Flight Number	From	To	STD	STA	Block Time	Mean Block Time	Block Buffer Time	Turn Time	Std. Turn Time
81	AP14	AP8	5.35	6.50	75	71	4		
82	AP8	AP14	7.10	8.25	75	67	8	20	20
83	AP14	AP17	8.55	11.00	125	115	10	30	30
84	AP17	AP14	12.20	14.35	135	125	10	80	40
85	AP14	AP15	15.10	17.35	145	135	10	35	30
86	AP15	AP14	18.15	20.40	145	141	4	40	40

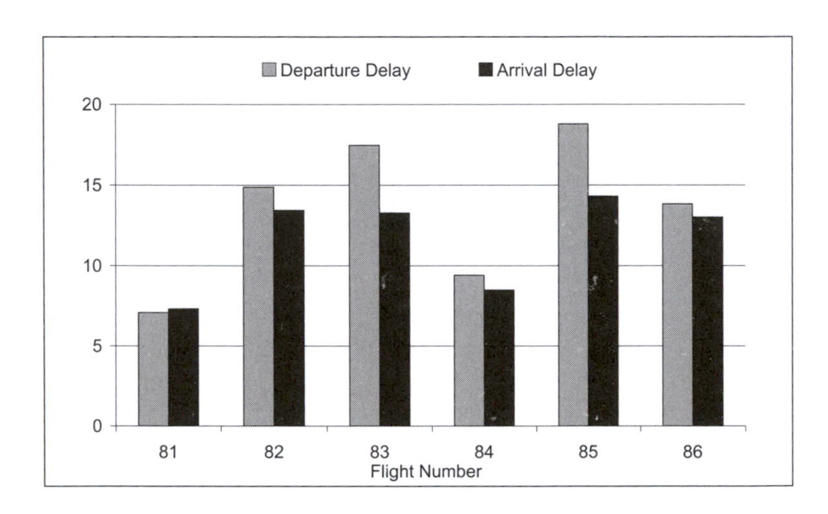

Figure 4.12 Actual departure/arrival delays of the study route (by A/C No. 8)

the departure delay of the second flight on the route (PP82). Delays escalated from seven minutes arrival delay by PP81 to fifteen minutes departure delay for PP82.

Delays were propagated along the route by this aircraft until the delays were controlled by the built-in long turnaround time in midday, which was 80 minutes for flight PP84. That was when the inbound delay by PP83 (14 minutes) was absorbed and controlled to the outbound delay by PP84 for only nine minutes. One would have thought that a long midday turnaround time with about 40 minutes extra buffer would have controlled delay propagation on the route after PP84 to a greater extent. However, the actual operation of this route revealed that ground disruptions to PP84 and delays due to crew changes (from Airline P's delay data) during the long break had further contributed to departure delays that propagated to the remaining two flights on the route, i.e. PP85 and PP86.

The fact that the last two flights on the route had the highest average delays was not unusual in airline operations. Since there is always a "natural buffer time" at the end of the day, e.g. after PP86, when aircraft are parked overnight and/or for maintenance checks on the ground at airports, accumulated delays by the last few flights in the evening will be naturally controlled. Hence, there is often little incentive for airlines to improve punctuality of late evening flights, unless delays to those flights are beyond tolerance or may cause serious disruptions to operations on the following morning, e.g. positioning aircraft overnight at some airports before curfew for next day operations. Failing to position aircraft back at certain airports before curfew may cause early morning disruptions to airline operations and this disruption scenario is among those serious cases that can cost airlines dearly.

4.5.2 The Enroute Simulation (ES) Model

Given the fact that an airline network comprises a number of routes, planned for all fleets to cover all flights, and a route is composed of a number of chronologically sequential flights linked by aircraft operations, we can easily extend the *Turnaround Simulation* (TS) model detailed in Section 3.4 of Chapter 3 to model airline operations on a network scale. The "operation of a flight" is defined as: *starting from the scheduled arrival time of the previous flight (if any) at the origin airport to the scheduled arrival time of this flight at the destination airport*. Thus, the "operation" includes two parts: ground operations (turnaround) at the origin airport and enroute operations between the origin and destination airports.

The TS model outlined earlier can be used to model the turnaround part of a flight operation, so we will develop an *Enroute Simulation* (ES) model in this section to model the remaining part of a flight operation, i.e. the enroute operation. Combining the TS model and the ES model recursively for a flight, we can model any individual flight operations. By "recursive", we mean that the outputs of the TS model such as departure time and delays, will serve as the inputs to the ES model; the ES model will then produce the simulated arrival time and delay of a study flight.

The "enroute operation" of a flight starts from pushing-back an aircraft at a gate (corresponding to the actual departure time of the flight), followed by taxiing out, runway queues (if any), take-off (wheel off), airborne aircraft operations (flying), landing (wheel on), taxiing in, and does not finish until parking at a gate at the destination airport. Hence, the enroute operation involves a number of operating activities and most of these activities are facilitated and managed by airports and air traffic management authorities. Since airlines have little control over the processes involved in enroute operations (with the exception of the actual flying of an aircraft), operational uncertainties and consequent delays are difficult for airlines to manage such as taxi delays, runway-end queues and air traffic congestions.

As discussed earlier in Section 2.3 of Chapter 2, the actual time of departure of flight f_{ij}, t_{ij}^D is a function of the actual arrival time of the inbound flight by the same aircraft, $t_{(i-1,j)}^A$ and the realised turnaround service time of f_{ij}, i.e. $\hat{h}_i$ as in (4.9) below. The relationship between the scheduled/actual departure time of a flight and the scheduled/actual arrival time can be modelled by (4.10), as discussed earlier in Section 2.5 of Chapter 2 and illustrated in Figure 2.7. The actual time of arrival of f_{ij}, denoted by t_{ij}^A is a function of the actual time of departure, t_{ij}^D and the realised flight time between Airport A and B, denoted by $\hat{k}_i$.

$$t_{ij}^D = t_{(i-1,j)}^A + \hat{h}_i \qquad (4.9)$$

$$t_{ij}^A = t_{ij}^D + \hat{k}_i \qquad (4.10)$$

Since all terms in both (4.9) and (4.10) are stochastic variables, we can extend the TS model which is represented by (4.9), (and also earlier by (3.11) in Section 3.4 of Chapter 3) and build the Enroute Simulation (ES) model based on (4.10).

4.5.3 The Network Simulation (NS) model

From equation (4.9) and (4.10), we can observe the analytically "recursive" relationship between the TS and ES model; the variable t_{ij}^D is on the left hand side of the TS model (4.9) as an output variable, and also on the right hand side of the ES model as an input variable (4.10). This recursive relationship also explains why delays propagate along routes due to aircraft operations and ground operational delays. If we extend this "recursive" relationship from the combined TS/ES model of one flight to another following the chronological order of flights on the same route, we can build a *Network Simulation* (NS) model that can model delay propagation on a route basis. The collation of all route-based delay propagation models of a network, i.e. the NS model, will be able to represent delay propagation in an airline network.

For instance, as illustrated earlier by the "aircraft routing layer" in Figure 4.3, the layered network diagram, we can build a TS and ES model for flight SS020 and recursively link it with the other sets of TS/ES models on the same route operated by the same aircraft such as flight SS030 and others. Combining all route-based TS/ES models, we can produce a NS model to study airline operations on a network scale. Since the NS model is based on routes, the delay propagation between flights on different routes (due to passenger connections and crew changes) is not explicitly modelled. Rather, this delay propagation across routes is modelled by stochastic disruptions in the TS model.

4.5.4 NS Model Application

The NS model was implemented by Monte Carlo simulation techniques. Two sets of data were required for such a simulation exercise: airline schedule data as "environment parameters" and past operational statistics as "Monte Carlo parameters". Airline P's schedule was used in the following case studies, and the required statistics and parameters for the NS model were derived from historical data. Two example sets of "environment parameters" are shown in Table 4.5, representing the efficiency of turnaround operations at two airports (coded as: Airport No. 14 and Airport No. 17) in Airline P's European network. It is seen from past records that turnaround operations at Airport No. 14 experienced more disruptions, especially in passenger processing.

Each airport modelled in the NS model of Airline P had a corresponding "profile" like those shown in Table 4.5, so as to represent the heterogeneous operating environments at different airports in a network. Parameters used in the NS model were also calibrated and verified with past operational data to ensure that the simulation environment created by the NS model was as close as possible to the real world situation. The schedule of Airline P in the NS model was executed by simulation 1,000 times (representing 1,000 days of operation) in order to limit simulation noises. Readers who would like a more thorough understanding of simulation modelling procedures and implementation techniques should consult books on modelling and simulation such as Fishman (1997).

To simulate how delays may propagate in Airline P's network, we tested three different scenarios as follows, and compared them with the base scenario, i.e. the current operation of Airline P's network.

- Scenario A: the base scenario (the current operation).
- Scenario B: disruption occurred to an early morning flight (PP121) and caused a ten-minute delay.
- Scenario C: disruption occurred to a midday flight (PP125) and caused a ten-minute delay.
- Scenario D: low turnaround efficiency at the base airport (Airport No. 1) of Airline P, causing ten-minute delays to those flights turned around at the base airport.

To test these scenarios, we used the route operated by aircraft No. 12 (shown in Table 4.6) which was planned for a shuttle service between Airport No. 7 and Airport No. 14 with a long visit to Airport No. 1 (the base airport of Airline P). We can see that the study route had a tight schedule with a short turnaround time at most airports. A 60-minute long turnaround time was planned for flight PP126 in midday as a fire-break for delay propagation. Little buffer time was planned for ground operations in addition to limited airborne buffer time planned for enroute operations.

Table 4.5 Example parameters used to simulate ground operations at AP14 and AP17

Parameters	Operation at Airport 14	Operation at Airport 17
Scheduled turnaround time	30 minutes	40 minutes
Expected turn time	20 minutes	30minutes
Cargo unloading (loading)	10 minutes (each)	15 minutes (each)
Pax deplane (enplane)	10 minutes (each)	15 minutes (each)
Standard deviation for cargo and passenger processing times	3–5 minutes	3–5 minutes
Probability to have:		
normal cargo unloading	90.9%	96.9%
cargo processing delays	0.1%	0.1%
ramp handling delays	9.0%	3.0%
Probability to have:		
normal cargo loading	84.9%	91.9%
cargo processing delays	0.1%	0.1%
ramp handling delays	9.0%	3.0%
passengers & baggage delays	6.0%	5.0%
Probability to have:		
normal pax boarding	70.0%	94.6%
crew delays	11.0%	0.4%
passenger delays	9.0%	1.0%
missing passengers	10.0%	4.0%
Probability to have:		
normal departure procedures	89.8%	98.4%
flight operation delays	0.6%	0.6%
departure process delays	9.0%	0.4%
weather delays	0.6%	0.6%

Table 4.6 Routing schedule of Aircraft No. 12

Flight Number	From	To	STD	STA	Block Time	Mean Block Time	Block Buffer Time	Turn Time	Std. Turn Time
121	AP7	AP14	5.40	6.55	75	69	6		
122	AP14	AP7	7.25	8.40	75	70	5	30	30
123	AP7	AP14	9.00	10.15	75	69	6	20	20
124	AP14	AP7	10.45	12.00	75	70	5	30	30
125	AP7	AP14	12.20	13.35	75	69	6	20	20
126	AP14	AP1	14.35	16.05	90	79	11	60	30
127	AP1	AP14	16.25	17.50	85	77	8	20	20
128	AP14	AP7	18.20	19.35	75	70	5	30	30
129	AP7	AP14	19.55	21.10	75	69	6	20	20

Scenario B was tested on the study route with a ten-minute disruption delay to the first flight on the route, namely PP121. The common wisdom in the industry is that delays to early morning flights are more likely to propagate in a network and cause further disruptions to other flights (AhmadBeygi et al. 2008a; Beatty et al. 1998; Wu 2006). Thus, Scenario B was aimed at testing this common wisdom and exploring the "sensitivity" of the study network. Results shown in Figure 4.13 revealed the comparison between the current operation (Scenario A) and Scenario B after the disturbance to PP121. We can see that delays propagated in Scenario A, growing from five minutes on average for PP121, to 16 minutes on average for PP125 and 20 minutes by end of the route, i.e. PP129. The long break time for PP126 broke the delay propagation chain by aircraft and controlled delays to a lower level. However, after the long break time, delays still propagated along the route. For Scenario B, we observed that delays to early morning flights caused further disruptions to this route. The increase of delays compared with Scenario A was mitigated by built-in buffer time in the route, although the trends towards delay growth between two scenarios were similar.

The results of Scenario C are shown in Figure 4.14 and are also compared with results of Scenario A earlier. We can see that disruptions to a midday flight (PP125) did cause further disturbance to the remaining flights on the route, although delays were effectively controlled by the built-in long break time on the route for PP126. Given the fact that only delays propagated via aircraft routing were explicitly modelled in this simulation model, the delay propagation effect was not seen as strong as expected. Other studies that considered delay sources other than aircraft routing, e.g. crewing in AhmadBeygi et al. (2008a) and Abdelghany et al. (2004) revealed a more complicated delay propagation mechanism that resembled what one would see in daily airline operations. In fact, the most complicated delay propagation scenario comes from considering the connections of passenger itineraries in addition to crewing and aircraft routing in an airline network. Following the delay propagation tree used by AhmadBeygi et al. (2008a), a delayed inbound flight may cause a binary split on the delay tree due to the separation of crew and aircraft. Moreover, the same delayed inbound flight may cause a five-split delay tree, more complex than the binary tree scenario, due to feeding passengers to five different outbound flights according to passenger itineraries.

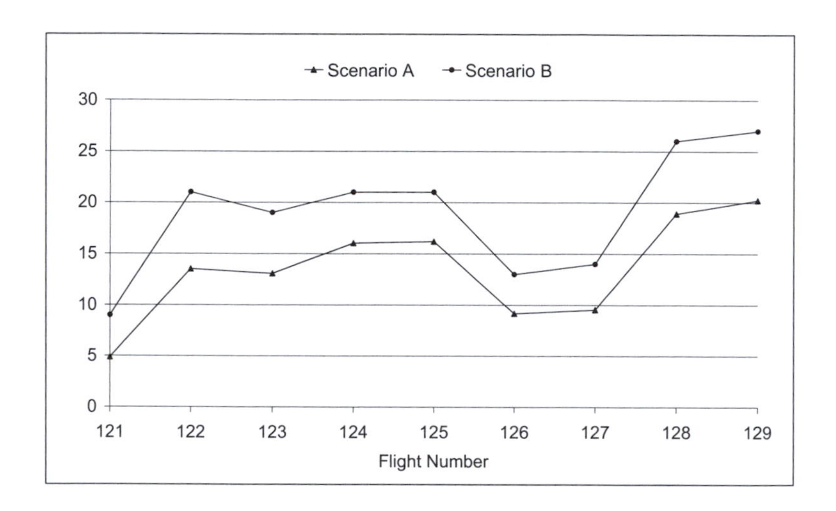

Figure 4.13 Delay propagation after PP121 (Scenario A and B)

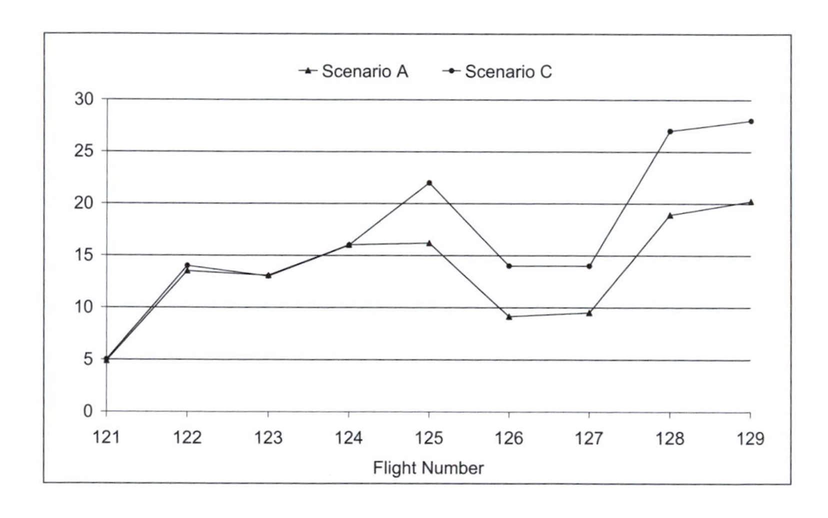

Figure 4.14 Delay propagation after PP125 (Scenario A and C)

Delays are often caused by airline operations at airports, besides those delays that are caused by air traffic system, airport capacity constraints and natural disruptions such as weather issues. To illustrate and explore the significance of ground operation control at airports, Scenario D was aimed at exploring the potential impact of low turnaround operational efficiency at the base airport (Airport No. 1) of the study network. Results of Scenario D are shown in Figure 4.15, illustrating how delays may propagate in the network due to low operational efficiency on the ground. From current operations, we can see that some routes (represented by aircraft numbers) performed poorly such as route No. 1, 2 and 11, while others had a satisfactory performance with delays within the band of ten to twenty minutes. After the disturbance caused by the low ground operational efficiency at Airport No. 1, the results of Scenario D showed that the impact was significant on a network scale. The disturbance to the ground operation at the base airport of Airline P disrupted 82 per cent of flights in the network and this resulted in a 65 per cent increase of total delays across the network in Scenario D compared with Scenario A.

This significant impact highlights the weakness of the network design of Airline P, due to the fact that a high portion of flights (82 per cent) of Airline P involved a visit to the base airport in an operational day. Hence, any disruptions to ground operations at Airport No. 1 had the potential to cause a significant impact on network integrity. By the same token, if Airline P allocated more resources to manage the performance at the base airport (for this example) or some other key airports in the network, then Airline P would be able to control network performance by the most cost-efficient way for its network, also due to the fact that controlling the operation at the base airport is equivalent to controlling 82 per cent of the operations of the network.

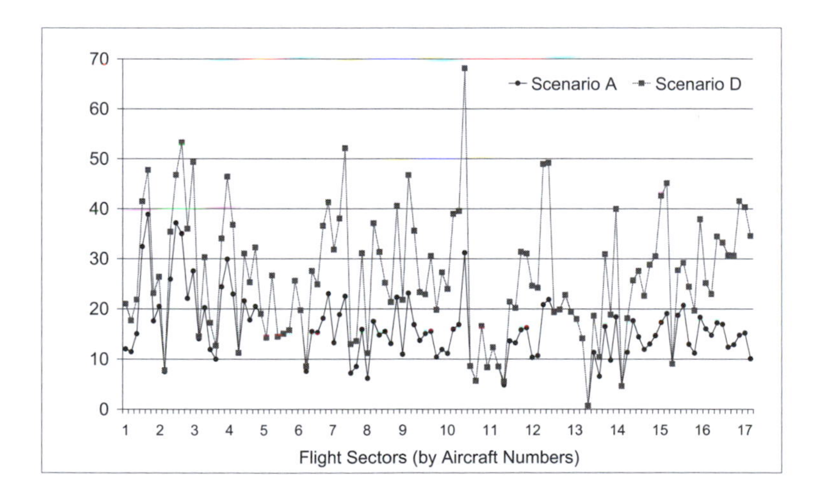

Figure 4.15 Impact of low efficiency in turnarounds at the base airport of P

Although not all airline networks have such a high concentration at the base airport as Airline P's, this case study highlights the implicit influence of airline network planning and resources allocation on daily airline operations. This conclusion echoes earlier suggestions that some operational issues are "caused" by the nature of the airline schedule and network planning and can be better solved by strategic schedule planning than operational disruption management means. Therefore, a good understanding of an airline network paves the way for a well-designed network and well-synchronised resource allocation, so as to achieve a high level of operational robustness.

4.6 Delay Propagation Modelling – Regression Models

Apart from the NS model, which uses simulation techniques in modelling delay propagation, an alternative approach is to model delay propagation in a network by statistical methods. Since there is a strong causal relationship between arrival delays via aircraft routing, ground delays due to aircraft turnaround disruptions, and departure delays, the development of a statistical model on the basis of strong causal relationships among delays and delay causes is warranted. Following the recursive mechanism we have discussed earlier while developing the NS model, we aim to develop a Departure Delay model and an Arrival Delay model for each individual flight in a network. Combining recursively the Departure and Arrival Delay models of multiple flights on the same route, we can create a *Route-based Delay Propagation* (RDP) model similar to the one we developed earlier for the NS model. A set of RDP models can be developed for each route of a network and the collation of the set of RDP models, called the *Network Delay Propagation* (NDP) model can be used to model delay propagation on a network scale.

4.6.1 The Network Delay Propagation (NDP) Model

Due to the causal relationship between arrival delays, delay causes and departure delays, the flight-based Departure Delay model has the following form as in (4.11). New notations and symbols introduced in this section are listed in the Appendix of this chapter.

$$d_{ij}^{D} = f\left(d_{(i-1,j)}^{A}, Delay\_Code\_Categories, p_{ij}, w_{ij}^{D}\right) \tag{4.11}$$

where d_{ij}^{D} denotes the departure delay of f_{ij}; $d_{(i-1,j)}^{A}$ denotes the inbound delay due to late inbound flight $f_{(i-1,j)}$. "Delay_Code_Categories" denotes a set of variables that represent major delay categories in the IATA delay coding system as shown in Table 4.7. There are twelve variables (V1~V12) in this set. "p_{ij}" represents the actual number of passengers carried by f_{ij}. "w_{ij}^{D}" is a binary (0/1) variable, derived from delay codes 71 to 77 (weather delays), with the exception of 72, which is used by the Arrival Delay Model for weather influence at the arrival airport. w_{ij}^{D} assumes the value of 1 if f_{ij} has delay records due to weather influences. This weather variable is not an "airport-level" variable, but an individual "flight-level" variable, because not all flights are subject to delays during a given weather condition. Hence, this variable provides the model with the flexibility to account only for those flights with w_{ij}^{D} delay records and not for all flights experiencing the same weather conditions.

Table 4.7 Independent variables in the flight-based Departure Delay model

Variables	Variable Selection Frequency	Mean Values of Coefficient	$(\mu,\sigma)^{*}$	Delay Occurrence Probability
TR Others (V1)	65%	1.17	(36.3, 30.3)	1.4%
TR Passenger And Baggage (V2)	91%	1.07	(5.8, 4.9)	8.9%
TR Cargo And Mail (V3)	34%	1.06	(6.9, 4.0)	0.6%
TR Aircraft And Ramp Handling (V4)	92%	1.33	(6.2, 4.5)	11.6%
TR Technical And Aircraft Equipment (V5)	88%	1.00	(28.3, 29.8)	3.8%
TR Damage To Aircraft (V6)	60%	1.13	(25.5, 23.2)	0.5%
TREDP Automated Equipment Failure (V7)	58%	0.75	(9.3, 7.1)	0.8%
TR Flight Ops And Crewing (V8)	83%	1.31	(6.3, 6.9)	5.9%
TR Air Traffic Flow Management (V9)	80%	1.33	(6.1, 5.6)	6.1%
TR Airport Government Authorities (V10)	76%	1.55	(2.9, 2.8)	7.7%
TR Miscellaneous (V11)	32%	1.43	(10.2, 7.6)	0.8%
TR Reactionary (V12)	84%	1.11	(12.0, 11.3)	6.5%
$d_{(i-1,j)}^{A}$ (inbound delay, V13)	88%	0.41	(14.9, 22.1)	43.4%
p_{ij} (passenger count, V14)	58%	0.12	–[†]	–
w_{ij}^{D} (weather delay at departure port, V15)	45%	25.74	–	0.9%
Model Intercept	–	-4.62	–	–

Note: * μ denotes the mean value; σ denotes the standard deviation. † Information not available due to confidentiality.

To avoid correlations between variables, which would imply a degree of dependency, correlation tests are necessary during the data preparation and model building stages. In correlation tests, those explanatory variables (V1~V15) of the Departure Delay Model in Table 4.7 were found to have little correlation among the variables. Originally, the inbound delay variable (V13, i.e. $d^A_{(i-1,j)}$) and the reactionary delay variable (V12) had a high correlation coefficient (0.56), because the original reactionary delay category included the delay code 93, i.e. delays due to "Aircraft Rotation". The separation of code 93 (as the Inbound Delay variable, i.e. V13) had reduced the correlation effectively. In addition, a major reason why we adopted the IATA framework in the regression model is that there is often little correlation among categories of delays, because one cannot assign the same delay to multiple sources.

For example, delayed ground handling due to weather belongs to the Weather factor and will not be included in the Aircraft and Ramp Handling category. Accordingly, the IATA framework has effectively kept the potential correlations among variables at the lowest level, while providing us with a rich set of operation-related variables in modelling delays on a flight basis. In addition, incorporating delay variables in the model helps significantly in explaining which other operational causes have a causal influence on delays, besides inbound/outbound delays. If we had only done the "naïve" regression model, using only departure delay and arrival delay, the results would have only reflected delay propagation, i.e. delays causing delays, without knowing how inbound delays become outbound delays after ground operations, i.e. the "mechanism" of delay propagation.

To model the delay propagation mechanism in flight operations, the multiple regression technique was applied to test the causal relationship between the arrival delay at the destination airport (the dependant variable), the departure delay at the origin airport, scheduled block time, airport congestion and weather influences. Since the recording of delay causes by airlines is carried out at the origin airport of a flight after its departure, delay causes due to enroute operations are often not available in the IATA coding system. Due to this limitation of data sources, statistical tests were conducted on Arrival Delay models in order to test whether those chosen model parameters (without detailed enroute delay causes) were adequate in representing the delay propagation mechanism of enroute flight operations. A flight-based Arrival Delay model has the form as in (4.12).

$$d^A_{ij} = f\left(d^D_{ij}, S^{BX}_{ij}, w^D_{ij}, w^A_{ij}, Peak^D, Peak^A, Hub^D, Hub^A\right) \tag{4.12}$$

where d^A_{ij} denotes the arrival delay of f_{ij} at the destination; d^D_{ij} denotes the departure delay of f_{ij} at the origin; S^{BX}_{ij} denotes the scheduled block time of f_{ij}. "w^D_{ij} / w^A_{ij}" are binary (0/1) variables, representing weather influence at the departure/arrival airport respectively. "$Peak^D/Peak^A$" are binary variables, indicating whether departures or arrivals were during peak hours. The "Peak hours" include the following time windows: 7–9am, 11am–2pm and 6–8pm. "Hub^D/Hub^A" are binary variables, denoting whether the departure or the arrival airport is a hub airport, a proxy of the likelihood of delays due to airport congestion.

The independent variables of the Arrival Delay model are listed in Table 4.8. Correlation tests were also conducted for the variables of the Arrival Delay model and results showed little correlation among variables. Indeed, departure delays occur before pushing back the aircraft from the gate. Accordingly, additional delays during taxiing or take-off queues due to weather or peak-hour airport congestion are not necessarily correlated to departure delays, because delay causes due to airport operations have little or no causal relationship with aircraft ground operations. Since the weather variables are on a flight basis, the correlation among variables via the "weather" factor was not found significant in the modelling process, implying that weather conditions did not affect all flights in the same way, unless in severe weather conditions.

The delay propagation mechanism between flights on the same route is modelled by the recursive relationship between the Departure Delay model and the Arrival Delay model. For flight f_{ij}, the flight-based delay propagation model can be derived by the recursive relationship between the Departure and Arrival Delay model, in which the departure delay variable (d_{ij}^{D}) appears on the left hand side of the arrival model, and the arrival delay variable ($d_{(i-1,j)}^{A}$) is on the right hand side of the departure model as seen in (4.13). For route j, the Route-based Delay Propagation model, RDP_j is derived by combining all flight-based models recursively as formulated by (4.13). For a network, the Network Delay Propagation (NDP) model is a collation of RDP models as in (4.14).

Table 4.8 Independent variables in the flight-based Arrival Delay model

Variables	Selection Frequency	Mean Values of Coefficient	(μ,σ)*	Delay Occurrence Probability
d_{ij}^{D} (departure delay)	100%	0.98	(13.7, 19.7)	57.1%
S_{ij}^{BX} (scheduled block time)	27%	-0.35	–	–
w_{ij}^{D} (departure airport weather)	28%	7.17	–	0.9%
w_{ij}^{A} (arrival airport weather)	19%	15.37	–	0.1%
Peak^{D} (peak-hour departure)	19%	-2.44	–	41.0%
Peak^{A} (peak-hour arrival)	22%	3.0	–	36.9%
Hub^{D} (departure hub airport)	11%	-12.2	–	60.6%
Hub^{A} (arrival hub airport)	0%	-8.8	–	60.9%
Model Intercept	–	3.63	–	–

Note: * μ denotes the mean value; σ denotes the standard deviation.

$$RDP_j = \bigcup_i \begin{cases} d_{ij}^D = f\left(d_{(i-1,j)}^A, Delay\_Code\_Categories, p_{ij}, w_{ij}^D\right) \\ d_{ij}^A = f\left(d_{ij}^D, S_{ij}^{BX}, w_{ij}^D, w_{ij}^A, Peak^D, Peak^A, Hub^D, Hub^A\right) \end{cases} \quad \forall i \in j \quad (4.13)$$

$$NDP = \bigcup_j RDP_j \quad (4.14)$$

4.6.2 NDP Model Application – Flight-Based Delay Propagation Models

A set of schedule and delay data from Airline S in 2004 was used to develop the NDP model of the network of this carrier. Another set of schedule and delay data in 2005 was used for cross validation of the NDP model built from the 2004 data. There were 563 flights recorded in 2004 schedule and approximately 90 routes which had been flown more than ten times in 2004. Since the data obtained were post-operation data, routes flown in 2004 were not necessarily those routes planned originally for aircraft routing (due to disruption management). Hence, a preliminary frequency screening maintained only those routes flown more than ten times in 2004, followed by a validation of the use of these routes in 2005. Sample sizes for each flight ranged from approximately 50 to 360 for a daily flight. In total, there were 563 flight-based models and 92 route-based RDP models in the NDP model of Airline S. Given the vast number of flight-based models, only one example result is shown in Table 4.9 to demonstrate the possible model forms and some model characteristics.

For this flight, both the Departure and Arrival Delay models showed a good fit to the recorded flight data with the adjusted R^2 values higher than 0.8. T tests and F tests also validated the statistical significance of the regression coefficients and the model itself at 95 per cent confidence level. The overall variable selection frequency during stepwise regression process for the Departure and Arrival Delay

Table 4.9 An example of the flight-based delay propagation model

Departure Delay model*	$d_{ij}^D = -0.49 + 1.03V1 + 1.05V2 + 1.04V3 + 1.04V4 + 1.01V5 + 1.00V6$ $+1.06V7 + 1.03V8 + 1.06V9 + 1.08V10 + 1.05V12 + 17.50w_{ij}^D$
(Remarks)	Residual standard error: 4.234 on **347** degrees of freedom Multiple R^2: 0.9512; **Adjusted R^2: 0.9495** F-statistic: **564** on 12 and 347 DF; P-value: < 2.2e-16
Arrival Delay model	$d_{ij}^A = 3.25 + 1.00d_{ij}^D + 5.56w_{ij}^D$
(Remarks)	Residual standard error: 8.361 on **361** degrees of freedom Multiple R^2: 0.8373; **Adjusted R^2: 0.8364** F-statistic: **928.7** on 2 and 361 DF; P-value: < 2.2e-16

Note: * Variable codes are based on Table 4.7 (given earlier).

model is given previously in Table 4.7 and Table 4.8. We can see that some ground operation variables revealed strong causal relationships with departure delays such as passenger and baggage handling (91 per cent selection), aircraft ramp handling (92 per cent) and inbound delays (88 per cent).

The mean values of the coefficients of variables of the flight-based Departure Delay models were computed and statistics on Table 4.7 showed that the mean coefficients were between 0.4 and 1.5, with the exception of the weather variable, w_{ij}^{D} . This implied that the impact of individual variables on the dependant variable was limited, at least for this specific data set. It was noted that the mean coefficient of "inbound delay" was about 0.41. When this information was cross-referenced with the mean delays of each ground operation variable, we found that the mean inbound delay of flights in 2004 was about 15 minutes with a delay probability of 43 per cent. This cross reference implied that although inbound delays occurred by a chance of 43 per cent in the network of Airline S in 2004, the contribution of inbound delays to the overall model was not as high as other variables. This could also be explained by the mean negative value of the model intercept as shown in Table 4.7, meaning that the scheduled ground time had enough of a buffer to offset the impact of inbound delays on outbound delays and the following propagated delays in the network. The high coefficient value of the "weather variable" showed that when delays occurred to a flight due to weather conditions, the impact on delays was often substantial. This inference is in line with what we have observed in daily airline operations.

On the other hand, the stepwise regression statistics from developing the Arrival Delay models in Table 4.8 showed that the variable, "departure delay" was always selected. Other variables such as the scheduled block time and weather conditions were selected with a frequency of only 27 per cent or lower. However, the mean values of the coefficients of variables in the Arrival Delay models in Table 4.8 revealed a mixed message. The impact of departure delays on arrival delays was almost proportional in quantity as the average coefficient was about 0.98. Flight block time, as expected had a negative impact on arrival delays. Regarding the weather influence, whenever weather caused delays, more influence was realised at the arrival airport than at the departure airport. The peak-hour factors and the hub-airport factors were not always statistically significant and were selected only 20 per cent of the time, or less, implying these factors were only significant for specific flights. This result also validated our earlier modelling decision of using the weather factors as a "flight-based" variable than an "airport-based" one, because not all flights that experienced a weather condition would incur delays in actual operations.

4.6.3 The RDP and NDP model

By recursively combining the flight-based Departure and Arrival Delay models of flights on a route according to (4.13), we are able to establish Route-based Delay Propagation (RDP) models. Since a RDP model comprises flight-based models recursively, the R^2 value of individual flight-based model influences the power of the RDP model in terms of explaining data variances, although not aggregately for the whole route. Due to the large number of individual RDP models tested in the network (totalling 92 routes), means and standard deviations of individual adjusted R^2 values of each flights on each route were calculated and shown in Figure 4.16. This overall view allows us to gain a high level understanding of the overall fitness of these RDP models to the data of the study airline.

We can see from Figure 4.16 that most Departure Delay models fit well the recorded sample data with high R^2 values in the range between 0.9 and 0.7. The standard deviations of R^2 values of the Departure models were below 0.2, showing the general consistency of model fitness on a flight basis. On the other hand, the mean values of the R^2 of Arrival Delay models ranged between 0.9 and 0.7, with some routes having lower mean R^2 values. Standard deviation values shown in Figure 4.16 also revealed that the cause of lower mean R^2 values for Arrival models was due to the variance of regression models among flights on specific routes. This implied that the overall fitness of the RDP model to the study data was good but further analysis may be required for those routes that had lower R^2 values in fitting delay data of Airline S.

4.6.4 Comparing the NS Model with the NDP Model

Compared with the NS model presented earlier, the biggest advantage of the NDP model is its analytical tractability and ease of development, although the task can be substantial for a large network and the development can be time-consuming. Once the required schedule and operation data are available, developing a NDP model based on statistical regression is not technically challenging for most analysts. Like other statistical models, the NDP model and in particular those

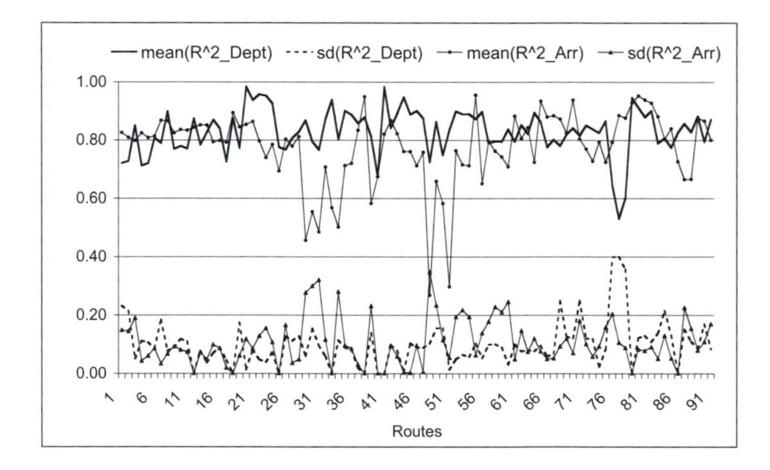

Figure 4.16 Overall fitness of RDP models

coefficients of regression models can reveal relevant messages that may have important implications for airline operations and network planning. For instance, one would be able to explore from flight-based delay propagation models the causal relationship between delays, schedules and potential delay causes during airline ground operations. Often, different flights reveal slightly different relationships, meaning that some disruptions may have higher impact on certain flights or on specific airline operations. Hence, statistical analysis is often used by airline analysts as both an analytical and a diagnostic tool that provides a starting point on studying delays, schedules and operations.

Interestingly, the biggest challenge in developing the NDP model was on data quality as well as result interpretation. Occasionally, airline analysts face a situation that analysis results do not reveal the expected causal relationships, or the statistical model does not have a good fit to airline data, e.g. low R^2 values for regression-based models. This implies that either the statistical model is not suitable for this analysis (on the basis that this methodology does not perform well), or for some cases, the quality of study data itself is not adequate for a sound statistical analysis.

On the other hand, the biggest advantage while developing the NS model was its capacity for the modelling of complex network structures and its flexibility on conducting scenario analyses. Indeed, the strength of a simulation-based model is that it can model very complex behaviours and systems with high precision, provided that we can understand and model smaller "components" that comprise the large and complex system. By modelling each "building block" of a complex system like the NS model, we are able to model a large airline network that involves complex airline operations at different airports and air traffic management dynamics. Surprises may emerge when a complex system is ran in a simulation model and one may be able to observe certain system "patterns" or "behaviours" under different simulating environments. In some cases, a complex emerging pattern in a large network may result from a simple causal relationship among building blocks in a large system such as the disease spreading models by cellular automata (such as: Green and Bransden 2006) and the emerging synchronisation of fireflies (Barabasi 2003).

The down side of the NS model is that like other simulation-based models, it often takes an investment of time to configure, test, and validate a simulation model before model implementation. This is especially true if the target system is large and/or complex in behaviour and it takes time to collect all required data in order to configure a simulation model. The NS model developed earlier did not explicitly include crew connections and passenger itineraries in the network. These two layers of an airline network can cause significant disruptions to airline operations, mainly due to the complex behaviour emerging from the interaction of these two layers with the layer of aircraft routing that we have explicitly modelled in the NS model. Interested readers can proceed in developing such a simulation model that explicitly considers all three layers of an airline network and the complex interacting behaviour among three layers. Surely, such a larger and more delicate

NS model will be useful and powerful in airline network analysis, although it may need a substantial quantity of resources and time to develop.

4.7 Summary

This chapter has expanded the discussion of airline operations from focusing on airline operations at airports in Chapters 2 and 3 to concentrating on operations in an airline network, taking a network perspective. The structure of an airline network was explored in depth including: the aircraft routing sub-network, crewing sub-network, and passenger itinerary sub-network. The complex interaction among sub-networks was examined and the emerging operational issues from network, scheduling and operational perspectives were discussed. A number of methodologies were presented that are informative and practical in measuring the complexity of airline networks for schedule planning purposes as well as for operational management purposes. A case study was given based on real airline data to demonstrate how network complexity measures could be used to study the implicit impact of airline scheduling (from the desire to pursue commercial benefits) on those emerging issues in operations that may eventually cost airlines.

In light of the network effects and operations, an operational issue, namely delay propagation, was further discussed with two models: a simulation-based NS model and a regression-based NDP model. Case studies were presented for both methodologies to show the implementation of delay propagation models in real cases, and how these models could improve airline schedule planning by providing feedback to the start of airline strategic schedule planning processes.

Until now, we have introduced readers with the concept of airline operations both on an individual airport level and on an enroute operation level, covering the "network" factor in airline scheduling and operations, the causes of delays, and the mechanism of delay propagation in an airline network. In the next chapter, we will shift the focus to day-to-day airline network management including schedule disruption management, delay management and delay reporting systems. A new scheduling concept, namely the *inherent delays* will be introduced in Chapter 5, based on those scheduling and operational concepts introduced in earlier chapters. This will further broaden readers' views on airline network design, operations, schedule optimisation techniques, and future trends of airline schedule planning that we will discuss later in Chapters 5 and 6.

Appendix
Notations and Symbols Introduced in Chapter 4

MCT	the minimum connection time required for passengers to connect to/from f_{ij}
CW	the inbound/outbound connection window in which passengers can connect from any flights to/from f_{ij} at Airport A /Airport B
d_{ij}^{ER}	the delays occurred after aircraft push-back at Airport A and before arriving at Airport B (mostly due to taxi, take-off and enroute delays)
M_i	is the number of potential inbound connecting flights to f_{ij} whose scheduled arrival times are within CW of f_{ij} (at Airport A)
N_i	is the number of potential outbound connecting flights after f_{ij} whose scheduled departure times are within CW of f_{ij} (at Airport B)
m_i	is the number of connecting flights inbound Airport A that caused connection delays to f_{ij} ($m_i \leq M_i$)
n_i	is the number of connecting flights outbound Airport B that received connection delays from f_{ij} ($n_i \leq N_i$)
d_i^m	is the total delays due to m_i disrupted inbound connections to f_{ij}
d_i^n	is the total delays propagated from f_{ij} to n_i disrupted outbound connections
d_j^m	is the total inbound delays of flights on route j
d_j^n	is the total outbound propagated delays of flights on route j
d^m	is the total inbound propagated delays in the network
d^n	is the total outbound propagated delays in the network

d^{op} is the total delays in the network due to ground operations (sum of d_{ij}^{op})

p_{ij} is the passenger number carried by f_{ij}

w_{ij}^{D} is a binary variable representing weather influence at the departure airport

w_{ij}^{A} is a binary variable representing weather influence at the arrival airport

Peak^{D} is a binary variable indicating peak-hour departures

Peak^{A} is a binary variable indicating peak-hour arrivals

Hub^{D} is a binary variable indicating whether the departure airport is a busy hub

Hub^{A} is a binary variable indicating whether the arrival airport is a busy hub

Chapter 5
Delay Management and Disruption Management[1]

This chapter covers both tactical and strategic aspects of airline operations, disruption management and methodological advances in schedule planning. Airline disruption management includes minor disruptions and major disruptions management, covering tactics widely used in the airline industry for mitigating the impacts of disruptions on airline operations and business in Section 5.1 of this chapter. The management of on-time performance (OTP) of an airline schedule is discussed in Section 5.2, covering the current practices of OTP management adopted by various carriers, the measurement of OTP by statistics, and the legal obligation of publishing and reporting OTP statistics by airlines in different countries. Examples and regulations drawn from FAA, Eurocontrol, Australia and various airlines are used to demonstrate OTP management and data reporting practices adopted in different regions of the world.

In Section 5.3, a case study is provided to demonstrate an OTP diagnosis of an example airline by studying its past OTP records, disruptions that occurred during operations, ground operational procedures, and potential impacts of scheduling strategies on operations and profitability. Given the close relationship between scheduling, stochastic disruptions and OTP targets, the concept of "inherent delays" of an airline network is introduced in Section 5.4. A case study demonstrates how this concept may help airlines design a reliable schedule and reduce the burden of real-time disruption management during operations. Schedule optimisation is further considered on a network scale in Section 5.5, together with the consideration of airline operations, operational uncertainties, schedule adjustments (flight re-timing) and network reliability in real operation. A case study here demonstrates the benefits of schedule optimisation by flight re-timing, including improvements on network reliability, as well as potential financial gains from robust operations.

5.1 Airline Disruption Management

Given the complexity of an airline network and the delicate synchronisation of resources among sub-networks of an airline network, there is a need to establish an Airline Operations Centre (AOC) to centrally manage operational disruptions that may occur in day-to-day operations and provide various units of an airline with

1 This chapter is partially based on the following publications: Wu (2005); (2006).

tactical disruption recovery plans (Clarke et al. 2000). Thus, the major function of an AOC, besides overseeing daily operations, is to co-ordinate recovery actions among different units within an airline such as crewing, ticketing and maintenance by "optimally" allocating limited operational resources, while minimising extra operating costs due to disruption recovery. By "optimally', it means that an airline may have certain disruption recovery policies that achieve the desired trade off between the goal of operation recovery and the goal of commercial interests, e.g. recovery with the least cost impact to the business.

Depending on the scale of disruptions, there are different recovery tactics. *Minor disruptions* refer to those ones that cause initial delays less than an hour, while *major disruptions* refer to the cases in which flights are delayed for longer than an hour. There are no specific rules as to why the "threshold" of delays adopted in this book is set to be one hour. From empirical evidence in the industry and observation of airline operations, one hour is roughly the threshold at which different recovery actions are taken, in particular when dealing with major disruptions. Often, it is seen that airlines adopt different thresholds for disruption recovery, reflecting the unique attributes of an airline network, e.g. network types, and certain in-house commercial and operational requirements such as the location of crew bases and aircraft maintenance schedules. For instance, an airline operating a point-to-point network would require different disruption recovery plans than an airline with a hubbing network, because of the way a point-to-point schedule is built (for both crewing and aircraft routing) and the nature of passenger flows in such a network. More details on schedule recovery will be discussed in the following sections.

5.1.1 Minor Disruptions and Recovery Tactics

As discussed earlier in Chapter 3, airline operations are subject to stochastic disruptions due to various reasons such as weather conditions, equipment readiness, passenger handling, and aircraft turnaround procedures. Most disrupting events only cause "minor" delays to certain operating tasks, and do not necessarily cause flight departure delays. For instance, inbound connecting passengers (and their bags) may be delayed due to late inbound flights. This lateness may cause disruption to passenger connection to some outbound flights. However, this does not necessarily imply that outbound flights with late connecting passengers and bags will be delayed. Departure delays only occur to a flight when the ready time to depart is later than the scheduled departure time. This has been mathematically modelled earlier in Section 3.4 of Chapter 3. Hence, minor delays may occur due to various operating tasks of an airline, but flight delays (departure/arrival) only occur when delays cannot be absorbed by the built-in schedule buffer time (if present).

Flight delays may propagate in an airline network due to resource connections. For instance, a minor departure delay, say of 40 minutes to a flight, may cause delay propagation to other flights at the destination airport where certain flights are to receive connecting passengers, goods, crew and even the aircraft itself.

Abdelghany et al. (2004) provided an example in which the departure delay of a flight may propagate to another 16 down-line flights due to crewing requirements and aircraft routing. Although passenger itinerary disruption was not discussed in Abdelghany et al.'s example, one can imagine how seriously passengers may be disrupted due to this scale of delay propagation.

However, it is noted that in a point-to-point network, passenger itineraries usually only span a single flight. Hence, in the similar minor disruption scenario, airlines which operate point-to-point networks have little (or none) issues with passenger re-accommodation due to missing connecting flights. Nevertheless, crewing and aircraft routing plans in a point-to-point network are still affected and for some cases, are affected more seriously due to the tight resource connections in a point-to-point network. Ironically, this feature of a point-to-point network is where most of its economical benefits and operational efficiency are gained, thus lowering the cost base of such a network (Gillen and Lall 2004).

Two major tasks of disruption management are to re-accommodate passengers whose travel itineraries have been disrupted, and to recover operational plans of crewing and aircraft routing. Under minor disruptions, passenger itineraries may be affected and the most commonly seen itinerary disruption for passengers in such a scenario is delays. Airlines usually offer delayed passengers compensation such as meal vouchers, if delays are due to the airline's own operations. A recent air passengers' right protection legislation in European Union also mandates that airlines must compensate passengers with meals, refreshments or accommodation if flight delays cause an overnight stay (European Union 2005).

Regarding the recovery of operational plans of crewing and aircraft routing by airlines, there are a number of tactics frequently adopted by airline operations controllers (Huang 1998). First, "doing-nothing" is always an option for minor disruptions. The rationale behind the "doing-nothing" option is that an airline schedule (more precisely, crewing and routing sub-networks) often contains some operational flexibility in the form of schedule buffer time. Buffer time exists in the ground time for turning around an aircraft as well as in the airborne time for flight operations. In a hubbing type of network, the buffer time between flights (allowing passenger connections, crew transfers and aircraft routing) is usually longer than in a point-to-point network operated by a low-cost carrier. With built-in flexibility in an airline schedule, "doing nothing" is always feasible, although passengers may be inconvenienced.

A recent study by AhmadBeygi et al. (2008a) on the potential of delay propagation in a hubbing network showed that minor disruptions can be easily absorbed by the built-in buffer time between flights, although this was more easily achieved for crewing and routing than for passengers. There are negative effects, however, due to flight delays and using this "doing-nothing" tactic for minor disruptions. Delays still cost airlines in such a situation and the loss of goodwill of delayed passengers may result in potential costs/losses in the future. Later on in Sections 5.4 and 5.5 of this chapter, we will discuss various scheduling methodologies that explore schedule flexibility and robustness, regarding the use of schedule buffer.

Second, when facing minor disruptions and expecting delays, airlines can alter operating procedures of ground operations and flight operations in order to "make up" possible delays and mitigate the impact due to disruptions. Two widely used tactics are:

1. speed up aircraft turnaround operations; and
2. speed up enroute flight operations.

Turnaround operations can be speeded up by allocating more ground staff (human resources), more equipment, and by the omission or delay of cargo loads. More resources (human or equipment) would shorten ground service time of some activities, e.g. bags loading/unloading, and catering loading/unloading. By omitting or delaying cargo loads to later flights to the same destination airport, the turnaround time of a delayed aircraft can be shortened. Sometimes, an airline may also delay connecting baggage to a later flight, if the cost of baggage delivery at the destination to passengers is less than the potential impact of flight delays. However, a flight must wait for connecting passengers, if connecting bags are already loaded due to security concerns, e.g. explosives or dangerous goods in bags (although this is rarely seen nowadays due to the advance of security screening technologies). Speeding up an aircraft during enroute operation can also save on delay time already accumulated. However, the cost of the extra fuel burnt due to higher cruising speeds must outweigh the potential costs due to delays.

Due to the nature of minor disruptions, flight cancellation is rarely used by operations controllers due to the high costs in passenger inconvenience and disruption to airline operating plans for crewing and aircraft routing. On the other hand, aircraft swap is a better choice in such a situation. In particular, swapping flights is easier if an airline operates a common fleet type (or a common fleet family) in a network such as all Boeing 737 or all Airbus A320, as adopted by many low-cost carriers. Aircraft swap can occur at an airport if an inbound flight is delayed and there are other aircraft at the same airport that are of the same or similar aircraft type. The swapping of aircraft between two routes would cut delay propagation immediately and has been a popular tactic for controlling minor delay propagation in a network.

This tactic is quite often deployed by low-cost airlines that operate a single fleet type of aircraft. Network carriers also deploy this tactic, but it often requires more collaborative actions from other units of an airline, due to more complicated fare (product) classes than low-cost carriers. Something to remember after swapping aircraft between routes is that an airline may also need to alter crewing plans and aircraft routing plans due to crewing constraints and aircraft maintenance schedules. Nonetheless, in a minor disruption situation, aircraft are often swapped back to the original routes later on the same day or the day after, so as to minimise costly alternation to crewing and maintenance schedules.

5.1.2 Major Disruptions and Recovery Tactics

Major disruptions to airline operations are mainly caused by inclement weather conditions (causing the reduction of airport runway capacity), aircraft mechanical faults, delays in airline operations, airline resources shortage, and rare incidents such as the incident of baggage system breakdown at Terminal 5 of London Heathrow Airport in March 2008 and security threats to airports/airlines. Dobbyn (2000) reported that 75 per cent of airline disruptions in the United States were caused by inclement weather conditions and the subsequent reduction in airport runway capacity. The impact of inclement weather and reduction of airport capacity on airline operations could be severe, if it occurs to a major airport or a hub airport of a specific airline (Ball et al. 2007). In such a situation, a hub airline is often forced to cancel many flights across the network in order to maintain the operation of "core" flight services.

Airlines' own operations may also cause disruptions. For instance, aircraft mechanical problems can cause severe disruption to a timetable, if the unavailability time of an aircraft is longer than two hours. If a serious mechanical situation occurs at an out-station where the airline has low service frequency (e.g. three flights per week), an aircraft can be grounded waiting for repair or parts for an extended period of time. This disrupts aircraft routing plans of the airline and major recovery actions must be taken to mitigate impacts. Airline resources shortage, e.g. crew shortage on the day of operation may also cause disruptions. This type of causes could be difficult to handle because they often emerge with a short lead-time to operations, limiting feasible recovery options an airline can deploy in a short response time window. Spare crew and even spare aircraft are critical for some airlines whose resources are heavily constrained by the nature of network design and/or collective crewing agreements.

Airline operations on the day are subject to unexpected major disruptions. For example, a delayed inbound feeder flight may cause a cascading impact at a hub airport to other outbound flights. For some cases, airlines need to "hold" departure flights in order to wait for connecting passengers, baggage and crew. If the "root" delay of the late inbound feeder flight is not minor (say longer than one hour), then delays to outbound flights may propagate in a network, compounded by disruptions to ground handling operations at the airport. Although the magnitude of delay propagation in this scenario significantly depends on the nature of network design and resources connection of an airline, delay propagation could develop significantly in some networks and seriously disrupt passenger itineraries. This case is more often seen in the operations of network carriers (or hubbing carriers), whose passengers have more complicated itineraries and on-line connections than those of low-cost carriers.

Schedule recovery under major disruptions is similar to that under minor disruptions, although the scale of recovery actions is larger, the impact on airline operating plans and passengers is larger, and some recovery options are preferred due to the differing effectiveness-cost trade-off relative to a short response time frame. Passenger re-accommodation is paramount for airlines during major disruptions (Clarke 2005). The European Union legislation on air passenger rights in 2005 mandates that airlines must provide passengers with meals, refreshments, and accommodation (overnight delays), if a flight is delayed by more than two hours (EU 2005). Apart from comforting delayed passengers in airport terminals, another facet of passenger re-accommodation is trying to transport those passengers by alternative means including ground transport (if feasible), flights by other airlines, flights by other routes between the same origin/destination airport of passenger itineraries, and flights scheduled at a later time.

Barnhart et al. (2002) modelled this problem as a multi-commodity network flow problem in order to minimise the total delay experienced by disrupted passengers. Clarke (2005) proposed the concept of passenger re-accommodation due to schedule changes by airlines, e.g. revenue management decisions, aiming at retaining the goodwill of "high-yield" passengers (first/business/frequent flyers) via priority re-accommodation. This is similar to the work by Bratu and Barnhart (2006) in which passenger itinerary recovery policies (e.g. frequent flyers recovery first etc) are enforced and flight operations are recovered with explicit consideration of passenger re-accommodation options simultaneously with aircraft and crew recovery.

Aircraft routing is affected due to schedule changes, e.g. flight delays and cancellations, or due to late aircraft changes such as changes/delays/aircraft swap due to technical problems. The goal of aircraft recovery is to limit the impact of disruption as well as minimise the disruption time. This task is often challenging for operations controllers of an airline, because any recovery plan must satisfy the following constraints, in addition to coping with the time pressure for quick recovery: 1) aircraft maintenance schedules; 2) airport operational constraints, e.g. curfew; and 3) aircraft flow "balance" at the end of disruption period. There are a few tactics in recovering aircraft routing back to normal operating schedules, including: delaying flights, flight cancellations, aircraft ferrying (flying an empty aircraft for repositioning purposes), diverting (flying to another airport), and swapping flights among aircraft routes.

Earlier work in aircraft recovery focused mostly on flight cancellations and/or flight delays, e.g. see Jarrah and Yu (1993) and Teodorovic and Stojkovic (1990). Often tactics are employed jointly as a part of the whole aircraft recovery plan. For instance, flight delays (imposed by airlines) can be deployed jointly with flight cancellations, or flight cancellations are executed together with aircraft ferrying (Yan and Yang 1996). Different considerations by airlines will lead to different recovery plans. In the integrated aircraft delay and cancellation model by Arguello et al. (1998), the objective was to minimise flight delays, cancellations and aircraft re-routing costs, due to repositioning aircraft in a network for aircraft balance requirement at the end of disruption period. Recent extensions in aircraft recovery research have also considered the limitation of airport slots (i.e. available time windows for take-off and landing) during disruption periods such as the work by Rosenberger et al. (2003) and Chang et al. (2001). Research has also gained significantly by jointly considering crewing recovery and passenger recovery requirements while working on aircraft recovery. The aforementioned work by Bratu and Barnhart (2006) considered disrupted passengers and crew pairing, while working on optimising aircraft recovery plans.

Since an airline network is a delicate synchronisation of passenger itineraries, aircraft routing and crewing sub-networks, disruptions to an aircraft routing sub-network and any recovery actions taken cannot avoid further disrupting the other two sub-networks. Crew disruption recovery is a highly complex issue for three reasons. First, crew costs usually constitute a major part of airline operating costs. Any disruptions to crewing would certainly cause extra crewing costs to an airline. Hence, minimising crew disruption costs is often an important goal of crew recovery. Second, crewing is subject to complex working rules (crew enterprise bargains) and safety regulations as discussed earlier in Section 4.2 of Chapter 4. These working rules and operational constraints must be respected in generating a feasible crew recovery plan. Third, crewing is highly synchronised with aircraft routing. This means that repositioning crew is as important in crew recovery plans as repositioning aircraft in aircraft recovery plans.

Crewing disruption, either caused by crew shortage or by aircraft routing, is usually dealt with by the following tactics:

1. deploy reserve crew;
2. deadhead (re-position) crew; and
3. rebuild/swap crew duties for disrupted flights/crew members.

As with other recovery problems for passengers and aircraft, crew recovery has multiple facets to consider such as minimising extra crew costs, minimising crew pairing disruptions and aiming to return to normal operation as soon as possible. Most crew recovery plans try to achieve a delicate balance between these facets, so as to achieve crew recovery with minimum financial and operational impacts on an airline. A typical crew recovery model is often multi-dimensional such as the work by Lettovsky et al. (2000) and Yu et al. (2003). Lettovsky et al. considered a crew recovery model with crew deadheading, work rules, flight coverage/cancellations, and costs associated with crew recovery, including passenger re-accommodation costs, and even the estimated loss of goodwill of disrupted passengers.

5.1.3 Choosing Recovery Plans and Considerations

Airline disruption management and schedule recovery is challenging due to the following reasons. First, an airline schedule is an ongoing operation with various resources connected with one another. Once a schedule is disrupted, any recovery plans must consider the recent operational history of aircraft (due to maintenance, flow balance and capacity supply), crew (due to crewing requirements and staffing levels), and passengers (due to the obligation to transport passengers). Second, a recovery plan often must be generated in a relatively short time frame with limited resources including reserved resources such as reserved crew and spare aircraft. Third, a recovery plan is often guided (and constrained as well) by an objective or multiple objectives in disruption management. They include the minimisation of passenger recovery costs, minimisation of disrupted time of airline operations (i.e. return to normal operation as soon as possible), and minimisation of airline costs due to disruptions. Often, some objectives conflict with each other due to the nature of resource connections and available recovery options. For instance, in order to recover airline operations to the normal stage as soon as possible, flight cancellations may be the desired tactic. However, this may not be the most economical choice (i.e. the cheapest) for recovery because it may cause heavy disruption to a wider population of passengers, crewing and aircraft routing.

For schedule recovery, there are always a number of options available to airlines. As to which option is eventually chosen and implemented to mitigate operational disruptions, decision-making may depend on the following considerations. First of all, the nature of disruption, i.e. the location, fleet types, time of day, and severity, determines the potential impact of a specific disrupting event on the whole network. The same aircraft breakdown for the same period of time at a hub airport of the airline is perhaps "easier" to deal with than one that happens at a spoke airport, because the hub airline often has more resources at the hub such as aircraft and maintenance facilities than at the spoke airport. A disruption that occurs late in the day may be more costly to an airline, because there are fewer options and lower chances of re-booking passengers on other flights (own flights or by other carriers). This would in turn imply that the airline needs to find hotel rooms to accommodate disrupted passengers.

Second, the network effect of disruption varies, depending on the "nature" of the disruption, the disruption time, network design, and most importantly resource connections, and passenger itineraries. For instance, if the flight disrupted is a key feeder flight by a network (hubbing) carrier, then the impact of that disruption could be severe "down the stream", causing propagation impacts on crewing, aircraft routing and passenger itineraries. In such a scenario, the airline would try to "contain" disruption propagation by swapping aircraft/crew, delaying some flights and re-booking passengers on other flights.

A similar disruption by a late inbound flight would trigger different recovery actions by a low-cost carrier (LCC). Since there are no on-line passenger connections, a LCC would prefer delaying flights which are disrupted by late aircraft and crew. If the delay becomes too long and there is little chance that delays can be "naturally contained" by the embedded buffer time in the network, a LCC would cancel flights, re-book passengers, re-route aircraft and re-organise crew. Flight cancellation is easier for a LCC because many LCC routes are run as shuttle services between two cities with high frequency. Hence, the cancellation of flight cycles, i.e. out and back between two cities, is easier for a LCC network, and it is also easier to re-book passengers on later flights, due to high flight frequency. The less complicated passenger itineraries, aircraft routing and crewing plans also make flight cancellations more economical and technically feasible for LCC disruption management. However, flight cancellations cause passenger inconvenience and itinerary disruption, which would affect the satisfaction of passengers in the long run.

Third, the cost of implementation is critical in choosing recovery plans. As discussed earlier, different recovery plans cause various levels of implementing costs. Ferrying an aircraft for repositioning is costly because there is usually little (nearly none) revenue from this operation. Yet, this option could be the one that could recover aircraft routing within the shortest time. Moreover, a recovery plan is often multi-dimensional and various recovery objectives may have trade-off effects among them. A straightforward trade-off is the aforementioned recovery time versus recovery costs. Apart from commercial losses due to disruptions and recovery plans, the loss of passenger goodwill can potentially be costly. This potential cost highly depends on how the frontline airline staff handle a disrupting event when it occurs. Monetary compensation such as meals and refreshments are not difficult to distribute by airline staff, but in most cases disrupted passengers would like to be more "involved" in the recovery process and be well informed of the situation and progress of itinerary recovery.

Fourth, schedule and network robustness considered in strategic schedule planning also influences how an airline chooses recovery plans. Since airline resources are expensive capital investments, the utilisation of resources (crew and aircraft) is essential in airline schedule planning stages, e.g. aircraft routing and crew pairing (Ball et al. 2007; Wu 2006). The consequence of high resource utilisation is the creation of a "tight" operational plan. This plan runs well during normal conditions, but may require significant and expensive recovery plans when disrupted. Hence, airlines have recently started exploring scheduling algorithms that can improve the operational "robustness" of a network by designing some sorts of "flexibility" in a schedule.

Rosenberger et al. (2004) proposed the concept of "short cycles" in fleet assignment to improve the "isolation" of a hub airport amid major disruptions to a hubbing carrier. Smith (2004) increased crew swap opportunities by building station "purity" in fleet assignment models, through limiting the number of fleet types accessing spoke airports. Station purity increased crew recovery options (lower recovery costs) due to common fleet types, although with higher planned costs. Yen and Birge (2006) considered various crew disruption scenarios at scheduling stages, and developed a stochastic programming model which was used to improve operating robustness of crewing. Wu (2006) explored the advantages of allocating different levels of schedule buffer times to flights in a network, aiming at improving the operational reliability of a network under minor disruptions. In light of this recent trend in airline scheduling, the concept of "schedule robustness" has emerged. More details of this schedule robustness concept will be given and further explored in Chapter 6, where we discuss recent advances and trends in airline scheduling.

5.2 On-Time Performance Measurement and Statistics Reporting

5.2.1 OTP Measurement and Statistics

The industry standard for the definition of an "on-time" departure/arrival is *a flight which departs/arrives less than 15 minutes after the scheduled departure or arrival time (off/on block time at a gate) shown in the published timetable of an airline or equivalently the time shown in the Computerised Reservation Systems (CRS).* This measurement is often measured by the percentage (%) of flights that are operated on time and this statistic is usually denoted by "D15" in the industry. D15 statistics can include both departure and arrival flights, depending on the purpose of measurement. In addition, "D0" is also used in the industry, denoting the percentage of flights that operate after the scheduled departure/arrival time as similarly defined by D15. D0 statistics, however, are not publicly available because airlines are not required to report D0 statistics to aviation authorities. Hence, D0 statistics are usually for in-house post-operation analysis only.

A commonly used OTP measurement statistic is the total delay time or equivalently the average delay time per flight. Depending on the purpose of this statistic, total delay time can be measured daily, weekly, monthly and yearly for individual flights, fleet types, operations (e.g. by regions, airports, and domestic or international operations), or for the whole network. For instance, the average delay time of a flight operated by A320 can be calculated and compared with that of another flight operated by the same aircraft type at a different airport. Airline analysts use this analysis to benchmark aircraft turnaround efficiency and procedures at different airports for future improvements. Earlier in Section 5.4 of Chapter 3, a few real examples from the industry were provided based on delay statistics. More will appear later in Section 5.3 of this chapter, when we use some industry data for a case analysis.

Another common OTP measurement statistic is to measure the probability of departure delays by delay causes. Earlier in Chapter 3, we discussed the framework of the IATA Delay Coding System and some in-house variations adopted by airlines. Based on the IATA system, airline analysts can generate the probability of delays according to flights, airports, fleet types and operations. A previous example given earlier in Section 3.4 of Chapter 3 highlighted the benefits of analysing such statistics. For instance, in that example it was found that there was a low chance (4 per cent) to encounter aircraft technical issues, but the impact in terms of delay time was extremely high (110 minutes on average). Although these delay probability statistics are useful for operations research in the industry, they are often not publicly available as operational data. Instead, various regulatory agencies, e.g. Eurocontrol in Europe and the U.S. Department of Transportation, require airlines to provide such data, but only publish statistics and reports in an aggregate form.

There are other OTP measurement statistics widely used by airlines, but mostly for in-house delay analysis and these are not publicly available. These statistics such as delay time variance and expected delay time, are derived from delay data and can provide invaluable information for delay analysis. For instance, the delay time variance of a flight can reveal the "shape" of the departure time distribution, allowing analysts to better understand the OTP of a specific flight. The expected delay time measure, on the other hand, factors in both delay time and delay probability. In particular, this statistic is useful when an analyst wants to further derive the expected "loss" due to flight delays by considering delay time, the unit cost of delay and delay probability (or delay distributions).

5.2.2 OTP Reporting Responsibility

Airlines around the world are required to report OTP data to the relevant governing agencies of aviation, depending on regulations of individual countries and governing agencies. In Australia, *Air Navigation Regulations 1947* set up the legal foundation upon which airlines or the operator of an aircraft registered under *Civil Aviation Regulations 1988* are required to return operational statistics (Australian Government 2003). These statistics include those data relating to: the aircraft, passengers or goods carried on the aircraft, work performed using the aircraft and flights made by the aircraft, and other statistical returns required by the Secretary.

The Bureau of Infrastructure, Transport and Regional Economics (BITRE) of the Australian Government is responsible for collecting and analysing transport related data. Since 2003, BITRE has co-ordinated a reporting scheme and established an agreement with major Australian airlines to report OTP statistics. The purpose of this OTP reporting scheme, like other governing agencies around the world, is to monitor airlines and overall industry performance, so air passengers can make informed judgements when planning air travel. OTP data on routes are published, if these routes are flown by two or more airlines where the passenger traffic averages over 8,000 passengers per month. Currently in 2008, there were 48 routes that met this definition in the domestic air network of Australia. OTP statistics and reports are available on the website of BITRE (see: http://www.bitre.gov.au).

In the U.S., Title 14 of the *Code of Federal Regulations* (CFR 14) regulates the functions of aviation related agencies such as the Federal Aviation Administration (FAA) and National Aeronautics and Space Administration (NASA) (Department of Transportation 2005). Part 234 in Chapter II of CFR 14 lays out the statistical return scheme for certain air carriers in order to provide air consumers with the information of airline service quality and performance. Airlines (or the reporting carriers) are required to submit data if they are certificated under 49 U.S.C. 41102, and they also account for at least one per cent of domestic scheduled-passenger revenues in the past 12 months ending March 31 of each year. In the *Air Travel Consumer Report* issued in December 2008 by the U.S. Department of Transportation (DoT), there were 18 airlines that met the reporting criteria and one carrier (Pinnacle Airlines) that reported data on a voluntary basis (Depeartment of Transportation 2008a).

The data provided by reporting air carriers in the U.S. are centrally collected and processed by the Bureau of Transportation Statistics (BTS) of DoT. Some airline operating data (mostly in aggregate form) are also available for the public via a web portal of BTS (see: http://www.bts.gov/programs/airline_information/). A large collection of analysis reports for the airline industry is available from BTS including: air carrier statistics, financial statistics, airline on-time performance, airport congestion status, airport tarmac delays and economic studies. Regarding OTP reports, the *Air Travel Consumer Report* produced by BTS is the major publication in the U.S. (Depeartment of Transportation 2008a). In the report, airline "service quality and performance" is measured by various criteria such as OTP, mishandled baggage, oversales, and consumer complaints. Moreover, OTP statistics are analysed and ranked among reported airlines by different criteria such as by carrier, by date, by time of day, by airport and by delay frequency. Given the increasing delays due to aircraft gate returns and waiting on airport tarmac, U.S. DoT expanded data collection in October 2008 to include those data when flights are cancelled, diverted or experience gate returns, providing air travellers with more information on tarmac delays and arrival delays (Depeartment of Transportation 2008b). Hence, the OTP statistics on tarmac delays are now available in the *Air Travel Consumer Report*.

In Europe, the governing agency of air safety and navigation is Eurocontrol that was funded in 1963 by six funding members based on an international (European) Convention, namely *EUROCONTROL International Convention relating to Cooperation for the Safety of Air Navigation* (Valtat 2007). The most recent revised Eurocontrol Convention in 1997 established two decision-making bodies, namely Eurocontrol Commissions and the Provisional Council, and one executive body, namely the Agency. Collecting aviation related data is the primary goal of Statistics and Forecast Service (STATFOR) within the Eurocontrol Agency. A STATFOR User Group guides the development and usage of statistical data collected by STATFOR, including methodological and practical aspects of developing statistical reports and forecasts.

Regarding traffic delay analysis, the Central Office for Delay Analysis (CODA) is the unit within the Eurocontrol Agency that is responsible for analysing traffic data and producing statistics and forecast reports. Reports are made available to the public on the internet based on data provided by airlines on a regular basis. In the recent 2007 annual OTP Digest, statistics were provided according to regions, traffic flows, airports (and pairs), delay causes, and average delays on a flight level (Eurocontrol 2007).

From the above three OTP reporting schemes, we can see that each scheme has its own focus. The BITRE scheme in Australia focuses mostly on reporting on-time performance of carriers and air routes. The *Air Travel Consumer Report* by BTS of DoT in the U.S. provides more detailed information on delays, on-time performance, carrier ranking on OTP, as well as other service quality surrogate measures such as mishandled baggage, flight cancellation and passenger complaints. On the other hand, the monthly/annual OTP digests produced by CODA of Eurocontrol focus more on air traffic delays and delay causes, as previously delays due to air traffic control and flow management were overwhelming in the European sky. By producing delay figures, OTP statistics and delay causes, stakeholders of Eurocontrol can have a more accurate picture of the evolution of air traffic flow management in the EU, as well as the essential initiative towards a "Single European Sky'.

5.3 Case Study – OTP Diagnosis

The following case study draws information from a carrier, denoted by Airline Z, to protect the identity of the carrier and the confidentiality of some sensitive data. The objective of this case study is to demonstrate how airline operational data can be used to preliminarily diagnose operational issues and help improve operational efficiency in future operations. The overall network OTP of Airline Z had been deteriorating gradually from 80 per cent (D15 OTP) in 2002 to 65 per cent in 2006. The decline of operational performance triggered internal reviews of flight procedures, ground operations and flight scheduling by Airline Z. Given the poor OTP performance in 2005 and 2006, more resources were required during disruption management, causing staggering extra operating costs and deteriorating passenger goodwill.

The following sections describe the steps that one would follow as a consultant in the airline industry to diagnose issues related to OTP. Although the following points may be useful in most real-world cases, one always needs to adapt these steps to unique circumstances, reflecting the availability of data from an airline and essentially, the objectives of such an exercise. When detailed operational data is available from a study airline, it always allows one to perform thorough analyses and produces invaluable results and recommendations for improving future operations and schedule planning.

5.3.1 The Network View – Delays in a Network and Delay Causes

The annual average OTP of Airline Z is shown in Figure 5.1. The dashed line shows clearly the decline of OTP between 2002 and 2006 that triggered the following analyses. On the same chart, the average daily aircraft utilisation of selected fleets showed the significant increase of daily utilisation across fleets. For instance, B737 utilisation improved from ten hours per day to nearly twelve hours per day in 2006, while B767 utilisation improved from 13 hours to 15 hours. The increase of aircraft utilisation was a part of a network expansion plan by Airline Z during 2002 and 2006, taking advantage of strong global economic growth and the increasing demand of air travel during that time. The nature of Airline Z's traffic was close to a weak hubbing network, in which regional feeder flights bring connecting passengers to other regional or long-haul inter-continental flights.

The high aircraft utilisation also came from the pressure of resources planning of Airline Z, in order to utilise expensive investments such as aircraft.

While the network of Airline Z expanded with more flights, more destinations and higher flight frequency, the overall network OTP suffered. This caused chain reactions in network operations, in particular delay propagation among flights and the staggering demand of resources required for disruption management. This phenomenon was commonly seen in the airline industry during 2003 and 2008, when the global economy was booming. High aircraft utilisation is often achieved by shortening aircraft turnaround times, flying longer sectors, flying overnight flights, and the scheduling of more sectors for individual aircraft. Since aircraft routing follows a series of connected flights, high aircraft utilisation is indeed a "double-edged sword". In normal operations, high aircraft utilisation brings significant cost gains as well as revenues. However, this plan can be easily disrupted, causing serious delay propagation in a network and deteriorating OTP as we have seen in the Airline Z's case.

5.3.2 Delay Propagation and Impacts on Operations

Detailed flight-level operational data were available for deeper investigation from Airline Z. After calculating delays and matching delay codes, total delay hours of each year were broken down into groups that were defined by group functions within Airline Z. Figure 5.2 shows the top five groups that caused the most delays during 2002 and 2006; they were: Airport/Civil Authority, Beyond Control, Reactionary, Passenger Services and Technical delays. Reactionary delays increased dramatically from the level in 2002 to that in 2006, nearly three fold. Technical issues also caused delays to increase by nearly two fold from 2002

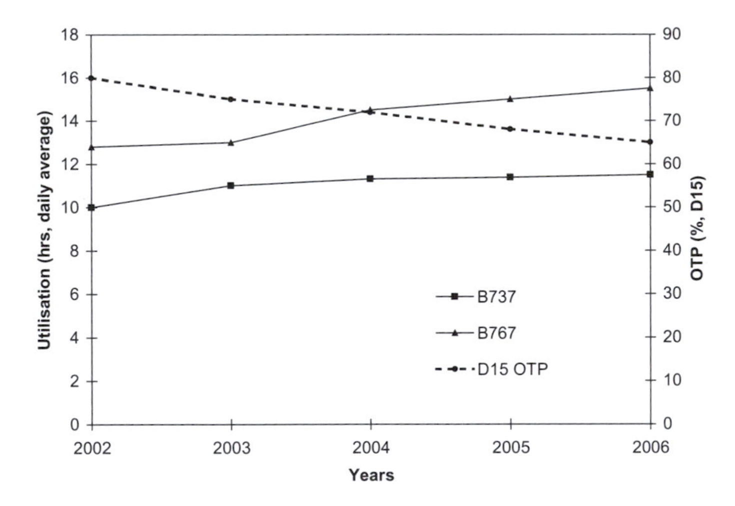

Figure 5.1 Aircraft utilisation of selected fleets and network OTP

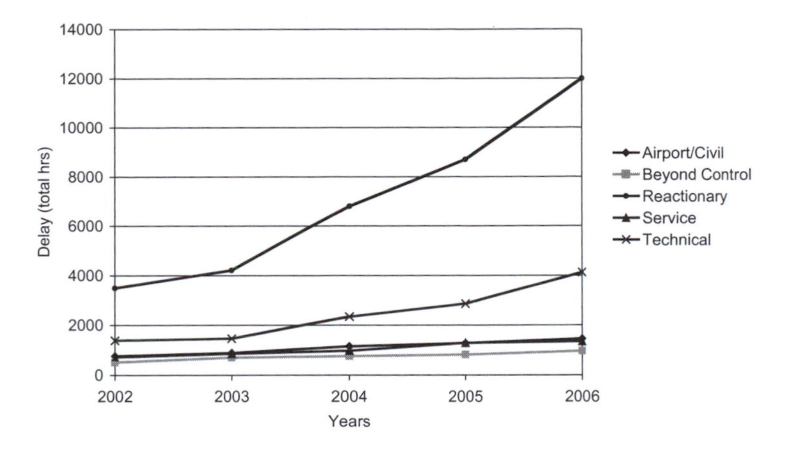

Figure 5.2 Delay ranking by ground service groups (top five only)

to 2006. It seemed that delay propagation in the network was a serious issue for Airline Z, but what were the root causes?

Statistics from available data showed that the probability of incurring technical issues was on average about 6 per cent, but there was a high 28 per cent chance of incurring delays due to reactionary causes. Although there was only a small chance of incurring technical delays, the consequence was that the average delay due to technical issues was as high as 132 minutes, nearly two hours for each incident on average. The impact of these technical delays to Airline Z was significant, and clearly the consequence was serious delay propagation (reactionary delays) from aircraft routing, crewing and passenger connections in the network.

A further break down of delay causes within the Reactionary group in Figure 5.3 reveals some hidden stories. We can see that delays due to RA (late inbound aircraft) went from 3,500 delay hours in 2002 to nearly 12,000 delay hours in 2006 (out of scale in Figure 5.3). While RA caused most of the reactionary delays, RL (connecting passenger load) and RO (reactionary ops control) delays also increased significantly, meaning more passengers missed connections or aircraft were delayed to wait for connecting passengers. Further evidence of delay propagation came from the increasing delays due to RO (reactionary ops control) and RC (late connecting crew from inbound flight). During operations, delays were more serious than we have seen from Figure 5.3, because the data we had for analysis was "post-operation" data. This means that the flight-level data was the operational "results" after disruption management of Airline Z; long delays had already been contained by disruption management tactics and this was reflected in the analysis data we have used here.

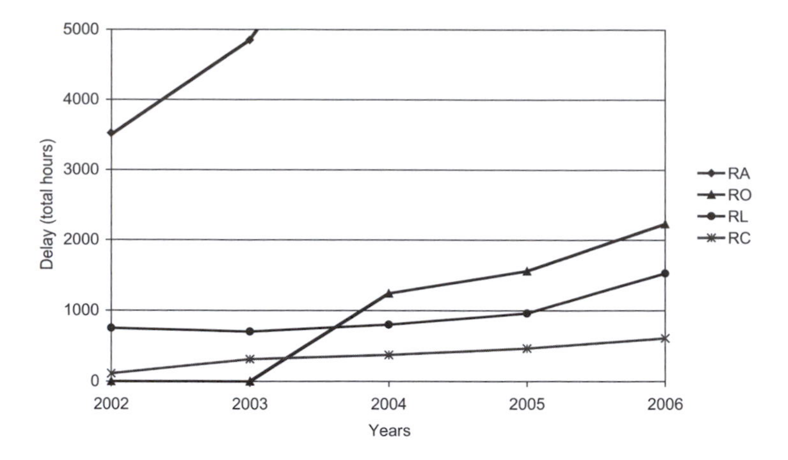

Figure 5.3 Delay causes breakdown for Reactionary group

5.3.3 Implications on Airline Operations and Schedule Planning

Among the top five ranked delay groups as shown in Figure 5.2, the Passenger Service group contributed equivalent amounts of delays to the Airport/Civil Authority group. The Passenger Services group covered work related to passenger services such as goods/catering loading/unloading during aircraft turnaround and passenger processing both on the airside and landside of an airport. This type of delays is more "controllable" and manageable by an airline by improving operating procedures, ground resources allocation (human and equipment) and operational efficiency. OTP data showed that Airline Z still needed to improve its operational procedures and efficiency in the area of Passenger Services.

Strategically speaking, we can diagnose Airline Z's operations by the following remarks. First, Airline Z had not had good control over "controllable delay causes", e.g. Passenger Services and Technical issues. As we can see from the ranking of the top five delay groups, three of them were controllable and manageable by the airline, the exceptions being delays due to Airport/Civil Authority and Beyond Control reasons. In fact, these two groups contributed far less delays to the network than the other three groups.

Second, high aircraft utilisation exacerbated the degradation of network operations by Airline Z. On one hand, "root" delays occurred due to poor passenger-related service procedures/efficiency, while technical issues caused significant disruption and delays to daily operations. Although the disruption management team of Airline Z could roll out some measures to ease the impact of disruptions, high aircraft utilisation left Airline Z with little room to manoeuvre but to implement costly schedule disruption mitigation actions, e.g. flight cancellations and re-scheduling.

Regarding strategic schedule planning, Airline Z has some options that can improve its operation in the future. Often, airline schedule planning takes limited feedback from past operations and simply builds new schedules by modifying last season's. This causes an airline to expand its network and operations based on a weak operational foundation because weak operational links are not addressed in the new timetable. In the particular case of Airline Z, an important goal in the future schedule planning is to improve schedule "flexibility" and operational robustness. Schedule flexibility can be achieved by allocating appropriate ground time for individual flights, considering the circumstances of ground operations (at individual airports), available resources for disruption management, passenger connections among flights, and rooms for improving ground operation efficiency.

Although allocating appropriate buffer time and re-time flights can reduce delay propagation and improve network robustness in operations, this strategy will unavoidably increase the use of schedule buffer time and reduce aircraft utilisation. However, compared with potential gains from decreasing delays, resources utilisation in delay handling and improving passenger goodwill, the path to improve schedule flexibility and operating robustness is worthwhile. More empirical evidence and theoretical background on this emerging scheduling concept, namely *Robust Airline Scheduling*, will be discussed in Chapter 6. Something to note here is that delays are "inherent" in any man-made system that operates a planned timetable such as a bus network or an airline network (Wu 2005). As long as available resources are limited, delays will always exist during operations because of stochastic disruptions. Seemingly unavoidable, some delays are indeed controllable and manageable, both strategically and tactically, especially for those potential delay causes that an airline has good control over.

5.4 Inherent Delays of an Airline Network

5.4.1 Fixed Schedules in a Stochastic Operating Environment

As introduced earlier in Chapter 1, the traditional procedure of airline schedule planning is partitioned into four sequential steps from schedule design (demand forecast), fleet assignment, aircraft routing to crew scheduling. Generally speaking, schedule optimisation objectives in various scheduling stages may include: minimising operating costs in aircraft routing, maximising profits in fleet assignment, maximising market penetration in schedule design, and maximising the utilisation of fleet and crew in routing and crew scheduling. Since the optimisation process tends to generate tight aircraft routing plans in order to maximise resource utilisation, schedule buffer times are usually embedded when solving aircraft routing and crew scheduling problems. The major function of the embedded buffer time in a schedule is to control small-scale (minor) delays and meanwhile maintain the required level of on-time performance (OTP) and operational reliability of schedules amid stochastic disruptions.

Since new airline schedules are published well ahead the new season starts, airlines have little flexibility in altering timetables during operations, except for minor alterations. Major and frequent timetable alternations would risk an airline losing airport slots and being not able to schedule flights at the same slot in the next season, due to the so-called "grandfather's right" and "use-it-or-lose-it" slot allocation policy in the industry. Given the nature of stochastic disruptions (either on a large or small scale) from daily flight operations, airlines indeed operate "fixed" schedules in an environment that is subject to stochastic disruptions and operational uncertainties. During normal operations, the schedule OTP would be close to the planned timetable and the potential issues of operating a fixed schedule can be disregarded. However, when schedule operation deviates from plans, airlines realise the consequences of stochastic disruptions, and the "gap" between the "expectation" of schedule planning and the "reality" of schedule execution.

5.4.2 Inherent Delays – Definition and Modelling

After finishing schedule planning, aircraft routing plans are generated for individual fleets. It is generally believed by industry schedulers that once aircraft routing plans are designed, there exists an *inherent OTP level* for a schedule, which reflects the schedule planning philosophy of an airline, e.g. the trade off between using buffer time to create flexibility (with higher planned costs) and maximising fleet utilisation (with lower planned costs but less flexibility). This explains why senior schedule planners can often "foresee" problematic flight connections and delay propagation in a draft schedule plan, when some flights are scheduled to follow certain flights in aircraft routing (Watterson and De Proost 1999). The concept of "inherent OTP" is illustrated in Figure 5.4. In ideal operations and "naïve" schedule planning, all delays are absorbed by ample buffer time; this scenario is labelled as the *Perfect Case*, meaning the no-delay case. The real-world operating scenario that is subject to stochastic disruptions and consequent delays is labelled as the *Reality Case*; this represents the situation we observe in daily airline operations. In-between the Perfect Case and the Reality Case on the right of Figure 5.4, there exists a *Dream Case* that reflects the inherent OTP level of a planned schedule.

Since it is not economically reasonable to pad a schedule sufficiently in order to fully eliminate delays, the Perfect Case hardly ever exists in the real world. Instead, we can only observe the Reality Case after schedule execution. The inherent OTP level of a schedule, i.e. the Dream Case is defined as *the implicit OTP "expectation" of an airline on future flight operations, given the "willingness" of an airline to design schedule flexibility with a higher planned cost (via buffer time ... etc), and considering potential stochastic disruptions and delays*. In other words, inherent OTP can be defined as the expected OTP level, which reflects the trade off in airline scheduling between allowing higher schedule flexibility (with higher planned costs) and potential impacts of stochastic disruptions (causing extra operating costs). Whenever the delay level after operation is higher than the inherently planned level, it means that there are higher impacts from abnormal operations than expected, and the designed schedule is not as robust as expected in such a situation. If the overall network delay level is frequently performing at a level higher than the inherent level, it means more resources are needed to improve schedule flexibility and robustness in order to mitigate operational disruptions and their impacts.

Since the inherent OTP level is not derived from operations but is conceptual, we need to develop a model to evaluate the inherent OTP level of an airline schedule. Given that an airline network is a time-varying network with stochastic service times of ground activities and stochastic disruptions in the system, we adopt the *Network Simulation* (NS) model described in Chapter 4 (see Section 4.5). The advantage of using a simulation approach is its capability and flexibility of mimicking a complex system with stochastic forces. Also, the NS model allows us to observe the emerging system patterns in operations under the interactions between the fixed flight schedule and stochastic disrupting events rising in operations.

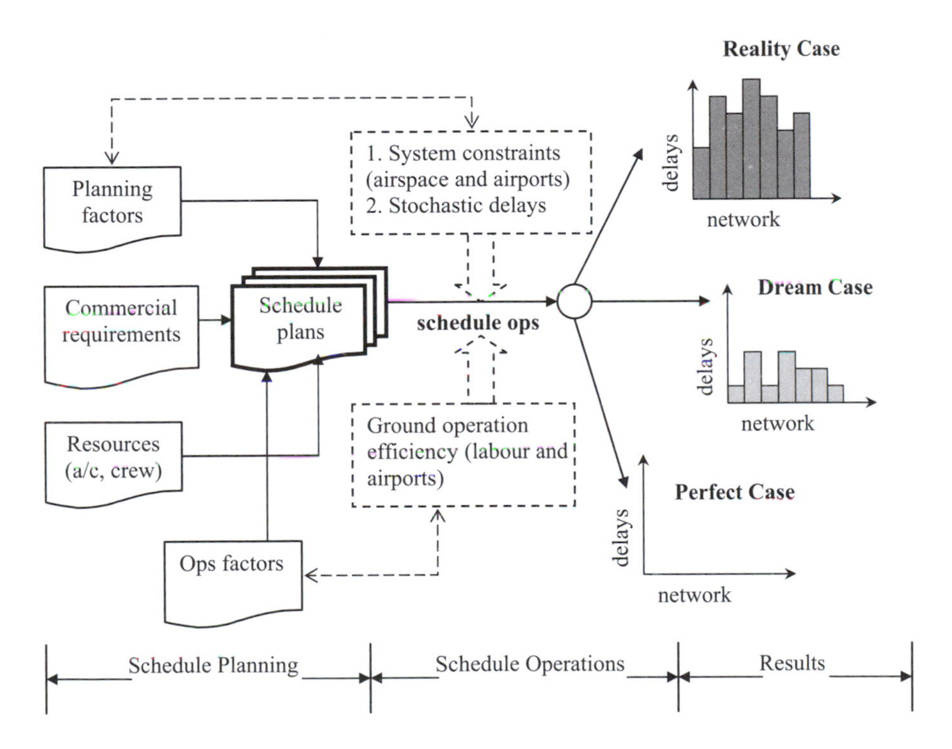

Figure 5.4 The conceptual model of inherent delays

5.4.3 Case Study – Exploring Inherent Delays of an Airline Network

Following the case study airline (denoted as Airline P) used earlier when we implemented the NS model in Chapter 4, we adopt the same case study framework here and focus on exploring the inherent OTP of Airline P's schedule. Parameters used in the NS model were obtained from the operating data of Airline P, including the planned timetable and statistics derived from that data. Some statistics served as "environment parameters" in the NS model, representing the operating environment and the occurrence of stochastic disruptions; other statistics served as "operating parameters', describing operational efficiency and certain operational characteristics such as the time needed to load cargo and bags.

The set of "environment" parameters was fixed after model verification and calibration with past operating data. The set of "operating" parameters, on the other hand, could be fixed after calibration, and could be used to conduct scenario studies like those we did in Chapter 4. Here we changed the set of operating parameters to reflect the "expectation" of airline operations. The "expectation" comes from the fact that most of airline/aircraft operations have standard operating procedures (also called SOPs in the industry). For instance, an airline would expect that if the aircraft turnaround operation at Airport X follows the SOPs of this type of aircraft, then the expected turnaround time of flights operated by this type of aircraft at Airport X would be Y minutes. Different SOPs and different expectations would lead to different sets of operating parameters for flights and airports in our case study.

To produce the inherent delays and OTP levels of Airline P, the calibrated set of operating parameters was modified to reflect different SOPs and expectations of individual flights and airports in Airline P's network. By running the original schedule with a new set of operating parameters, we were able to generate the inherent delays of Airline P as shown in Figure 5.5. This scenario was labelled as "Scenario E', continuing from the scenario study presented in Section 4.5 of Chapter 4. Hence, Scenario A represented the real delays after the operation of Airline P.

When real delays of individual flights are compared with simulated delays (i.e. Scenario E, representing the "inherent delays" and the Dream Case) in Figure 5.5, we can find significant differences between the two, from a few routes such as those by aircraft No. 2, 3, 5 and 10. This implies that delays in real-world operations were significantly higher than the "inherent level" expected after schedule planning. The inherent delay curve also reveals some insights. We can see from Figure 5.5 that Airline P's schedule plan was not expected to have a consistently low level of delays across the network as seen from the levels of inherent delays; some flights were "expected" to have delays between 10–15 minutes, but some flights were not. This means that across the network of Airline P, some flights had more exposure to disruptions and delays (perhaps due to higher uncertainties in operations) and this may cause delay propagation to other flights, if the delays could not be contained by the original flights that were expected to have low delay exposure. To further investigate the causes for varying inherent delays, the routing plan of an aircraft (No. 8) was chosen for further examination, representing a routing plan with medium operational reliability.

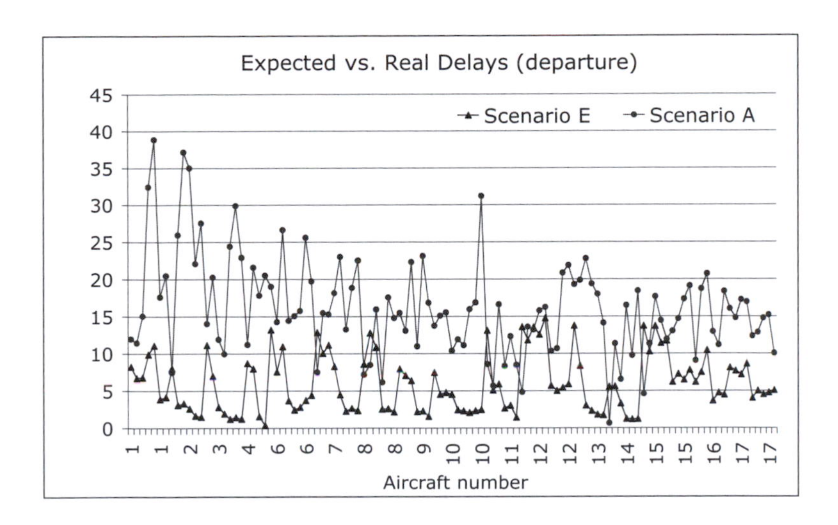

Figure 5.5 Comparison between the Reality Case and the Dream Case

Simulation results of Aircraft No. 8 are shown in Figure 5.6 and 5.7 (from Scenario E) together with operating delays from past punctuality data (Scenario A). Results showed that real operating delays of early flight segments, e.g. flight PP81 and PP82 were close to expectation, but later segments, e.g. PP84, PP85, and PP86 had higher delays than expected. A further investigation of the gaps between Scenario A and E showed that more airborne buffer time was planned for the later segments, resulting in a low level of expected delays in Scenario E. The routing plan by this aircraft is shown in Table 5.1 and we can find that a long turnaround time (80 minutes) was scheduled for PP84 at AP17 (the departure airport). The scheduled long turnaround time (with 40-minute buffer time) by Airline P was designed to absorb propagated delays from earlier segments. We can see that in both scenarios the long midday buffer time was effective in controlling delay propagation on this route.

Although the average departure delay of PP84 was less well controlled than earlier segments, delays from later flights of that route still increased significantly. Table 5.1 also showed that the planned aircraft turnaround time for most flights in this route was just sufficient for standard turnaround operations, meaning that any minor operating disruption could easily cause flight delays. As a consequence, delays were likely to propagate via aircraft routing and were not likely to be controlled by the schedule; this was exactly what Airline P observed in operations. The midday long buffer time of PP84 effectively reduced departure delays of PP84 and broke the propagation chain along this route. Any further delays to PP85 and PP86 were likely to originate from their own turnaround operations, possibly due to the fact that these two flights were scheduled to operate during the busy hours of a day.

We can also see from those two earlier Figures (Figure 5.6 and 5.7) that the developing trend of delays in Scenario E (the Dream Case) was consistent with the one in Scenario A (the Reality Case), but had lower delay levels. This implied that the operation of this route (including these flights) did not perform as well as expected from the perspective of schedule planning. OTP results (D15 OTP) from these two cases are compared in Figure 5.8. We can see that real operations of early segments had better OTP than expected, while those later flights performed relatively poorly, compared with the expected inherent OTP levels.

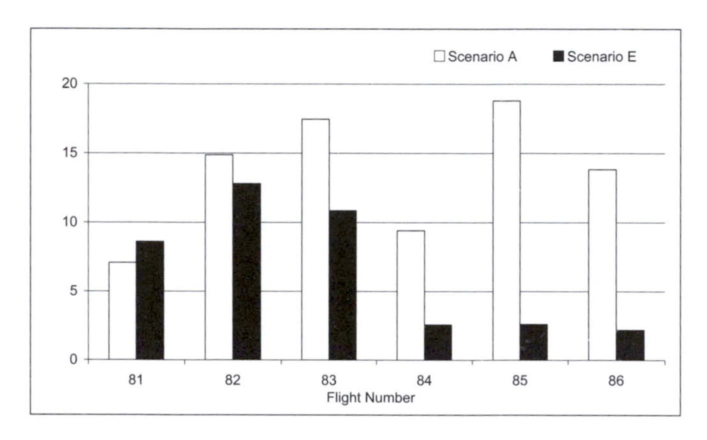

Figure 5.6 Simulation results of aircraft No. 8 (departure delays)

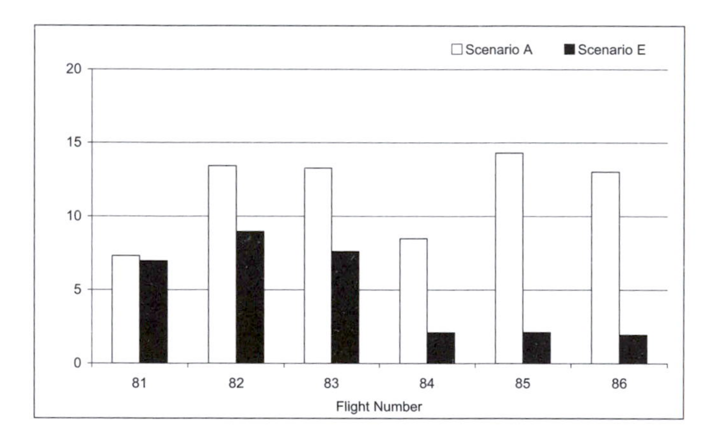

Figure 5.7 Simulation results of aircraft No. 8 (arrival delays)

Table 5.1 Routing schedule of Aircraft No. 8

Flight Number	From	To	STD	STA	Block Time	Mean Block Time	Block Buffer Time	Turn Time	Std. Turn Time
81	AP14	AP8	5.35	6.50	75	71	4		
82	AP8	AP14	7.10	8.25	75	67	8	20	20
83	AP14	AP17	8.55	11.00	125	115	10	30	30
84	AP17	AP14	12.20	14.35	135	125	10	80	40
85	AP14	AP15	15.10	17.35	145	135	10	35	30
86	AP15	AP14	18.15	20.40	145	141	4	40	40

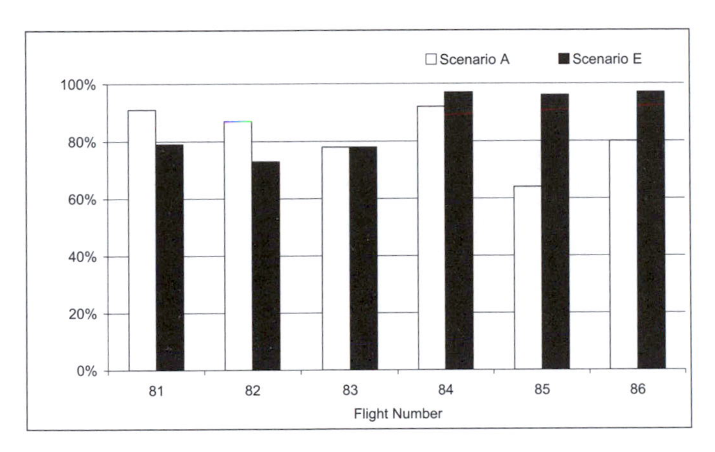

Figure 5.8 Simulation results of aircraft No. 8 (D15 OTP)

5.5 Schedule Optimisation and Network Reliability

As discussed earlier in Section 5.1, airline operations are subject to stochastic disruptions. Depending on the scale of disrupting events and the potential impact on the network, an airline takes actions to mitigate impacts or take no actions and hence, allows the network to naturally absorb delay propagation. The issue of operational reliability of airline schedules has attracted some attention recently in the industry. Further discussion of this emerging trend in airline scheduling will be covered in Chapter 6. The focus of this section is on defining schedule robustness, operational reliability and providing a scheduling methodology to achieve the desired reliability via strategic schedule planning.

5.5.1 Operational Reliability and Strategic Schedule Planning

Schedule *reliability* (also known as "*schedule robustness'*) is defined as both a concept and a measure that describes the "degree" to which an airline schedule performs in actual operation, compared with the original plan. We have discussed in earlier sections about how the actual airline schedule operation is subject to stochastic disruptions (i.e. the Reality Case) and how the planned schedule may lead to the expected schedule performance (i.e. the Dream Case or the inherent status), given those considerations in schedule planning. Hence, according to the definition of schedule reliability, the purpose of this measure is to evaluate the "gap" in operations between the actual operations (Reality Case) and the expected operating results, i.e. the Dream Case. The function of the inherent delays of a schedule now serves as a *benchmark* in evaluating actual schedule operations against what an airline would expect from its strategic schedule planning and operations.

To quantify schedule reliability, a set of schedule reliability indices is developed according to the availability of airline data and our definition of reliability. The *schedule reliability* of flight *i* on route *j* (denoted as flight f_{ij}) is defined as *the ratio between the inherent delays generated from the Dream Case and the actual delays from the Reality Case* as formulated by (5.1) below. Symbols and notations newly introduced in this chapter are listed in the Appendix of this chapter.

$$R_{ij}^{D} = \frac{{}_{e}d_{ij}^{D}}{d_{ij}^{D}}, \qquad R_{ij}^{A} = \frac{{}_{e}d_{ij}^{A}}{d_{ij}^{A}}, \qquad R_{ij} = \frac{({}_{e}d_{ij}^{D} + {}_{e}d_{ij}^{A})}{(d_{ij}^{D} + d_{ij}^{A})} \tag{5.1}$$

where R_{ij}^{D} / R_{ij}^{A} denotes the departure/arrival reliability of flight f_{ij} respectively; ${}_{e}d_{ij}^{D}$ / ${}_{e}d_{ij}^{A}$ represents the expected departure/arrival delay of f_{ij}, while d_{ij}^{D} and d_{ij}^{A} the actual departure and arrival delay of f_{ij}. Hence, R_{ij} is used to evaluate the overall operational reliability of flight f_{ij} according to (5.1).

The concept behind this schedule reliability index is to benchmark actual schedule delays against the inherent delays so airlines can evaluate how close the actual schedule operation is compared to the expected delay levels in schedule planning. The schedule reliability index reflects both the schedule planning philosophy, e.g. the willingness to buffer schedule, and the situation in which real operations are conducted by the airline, e.g. ground operational efficiency. For instance, if R_{ij}^{D} has a value larger than 100 per cent, it means that the actual departure delays of flight f_{ij} (d_{ij}^{D}) are less than the inherent delays (${}_{e}d_{ij}^{D}$). It implies that the actual operation outperforms the expectation and the scheduled ground time may even be shortened to save aircraft time, if needed. However, quite often we see R_{ij}^{D} has a value less than 100 per cent, meaning that the actual delay level is higher than the expected one. This may result from inadequate buffer time for turnarounds, delay propagation in aircraft routing, less robust turnaround operations, or excessive disruptions occurring in actual operations.

Following the same reasoning, the reliability of aircraft route *j* is modelled in (5.2), where R_j represents the reliability of route *j*. The network reliability, denoted by R_{net} in (5.3) is used to evaluate the network-wide reliability of a schedule. Reliability indices developed here can be used tactically to measure and benchmark the performance of individual flights such as by R_{ij}, or strategically to measure the overall reliability of a network system such as by R_j and R_{net}.

$$R_j = \frac{\sum_{i \in j} ({}_e d_{ij}^D + {}_e d_{ij}^A)}{\sum_{i \in j} (d_{ij}^D + d_{ij}^A)} \qquad \forall i \in j \tag{5.2}$$

$$R_{net} = \frac{\sum_j \sum_i ({}_e d_{ij}^D + {}_e d_{ij}^A)}{\sum_j \sum_i (d_{ij}^D + d_{ij}^A)} \qquad \forall i \in j \quad \forall j \in F \tag{5.3}$$

5.5.2 Schedule Optimisation – Flight Re-timing Model

Currently, the integer programming approach is widely used by airlines to solve aircraft routing problems. However, given that integer programming is not an operation-oriented methodology, optimisation results tend to produce tight schedules under cost minimisation objectives without thorough consideration of delay propagation effects in actual network operations. Very often it is seen from airline operations that particular flights need more turnaround time than others due to the efficiency of turnaround at certain airports, connecting passengers among flights and so forth. The integer-programming approach for optimising aircraft routing may also cause other issues in actual operations such as delay propagation, frequent delays to specific flights and tight turnarounds. Experienced airline schedulers often manually fine-tune schedules because some aspects of airline operations are not explicitly considered in conventional optimisation models.

Based on the current practice of aircraft routing optimisation and schedule adjustment procedures in the industry, a *Sequential Optimisation* algorithm is developed aiming at supplementing existing integer programming algorithms widely adopted in proprietary schedule optimisation packages used by airlines. The goal of Sequential Optimisation is to improve the efficiency and effectiveness of the manual schedule tuning process, so as to improve schedule robustness and reliability in operations.

After initial aircraft routing optimisation, a routing plan is generated which consists of J routing patterns (operated by J aircraft) to cover F flights in a network. Based on the "schedule reliability" defined earlier, the goal to improve the reliability of flight f_{ij} is to control the actual delay levels as close as possible to the benchmark standard, i.e. the levels of inherent delays. Hence, to optimise the operational reliability of a schedule, the given aircraft routing patterns are further relaxed by allocating extra schedule buffer time in aircraft ground time, i.e. flight re-timing (Lan et al. 2006; Lee et al. 2007; Sandhu 2007; Wu 2006). Since buffer time is an expensive item for airlines, the use of buffer time in the schedule relaxation algorithm shall be constrained to achieve a chosen performance target, which is measured by delay time in this model. The reliability target, denoted by R_{tar}, can be chosen arbitrarily by an airline as the schedule operation target. According to the definition of the reliability index given earlier, the resulting target departure delays for flight f_{ij} is:

$$_{tar}d_{ij}^D = \frac{_{e}d_{ij}^D}{R_{tar}} \tag{5.4}$$

Accordingly, the schedule adjustment for flight f_{ij}, denoted by Δs_{ij}^D, can be expressed by (5.5) below, where d_{ij}^D denotes the current delays of flight f_{ij}, or the simulated delays of f_{ij}, if any preceding flights in the same route as f_{ij} are subject to flight re-timing. It is noted that when Δs_{ij}^D has a positive value, it implies that the current delay level of flight f_{ij} is higher than the target level, and a negative value implies the reverse. This also implies that by flight re-timing an airline may save excessive aircraft time from those flights which have negative Δs_{ij}^D, and use the saved aircraft time on those flights which have positive Δs_{ij}^D.

$$\Delta s_{ij}^D = d_{ij}^D - {}_{tar}d_{ij}^D \tag{5.5}$$

Therefore, the total available schedule adjustment time for route *j* (denoted by Δs_j^D) can be expressed by (5.6).

$$\Delta s_j^D = \sum_i \Delta s_{ij}^D \tag{5.6}$$

Since flight delays and flight punctuality both propagate along aircraft routing in a network, the relaxation of the ground time of earlier flights in route *j* will reduce the likelihood of delays of following flights in the same route. In other words, if flight f_{ij} receives schedule adjustment Δs_{ij}^D (where $\Delta s_{ij}^D > 0$), the potential delay propagation from f_{ij} to the next flight $f_{i+1,j}$ will be reduced. Accordingly, the initial schedule adjustment for flight $f_{i+1,j}$, denoted by $\Delta s_{i+1,j}^D$, would be more than necessary to achieve the target reliability, R_{tar}, due to *punctuality propagation*. Following (5.5) earlier, $(d_{i+1,j}^D)'$ of $f_{i+1,j}$ from the simulation is now less than its original value ($d_{i+1,j}^D$), due to the benefit of "punctuality propagation" from f_{ij}. Hence the required schedule adjustment time for $f_{i+1,j}$, i.e. $\Delta s_{i+1,j}^D$, will be less than the amount required before the adjustment of f_{ij}. This "punctuality propagation" concept will be further discussed in the up-coming case study.

Given the nature of delay/punctuality propagation through aircraft routing, the Sequential Optimisation algorithm is developed for flight re-timing, and is expressed by (5.7) as follows:

To minimise: $$\sum_j \sum_i d_{ij}^D \tag{5.7}$$

Subject to: $$\{d_{ij}^D \leq 15\} \vee \{d_{ij}^D \leq {}_{tar}d_{ij}^D\} \tag{5.8}$$

where the objective of the Sequential Optimisation algorithm is to minimise departure delays as in (5.7). This is equivalent to minimising flight arrival delays, assuming that flight block time is not changed and that the operating environment does not change significantly. The only constraint in this flight re-timing model is that the estimated delay of individual flights after schedule adjustment is under 15 minutes (which is the industry delay threshold), or under the chosen target delay level (${}_{tar}d_{ij}^D$), whichever requires less buffer time. To implement the flight re-timing model, the Sequential Optimisation algorithm can be conducted as follows:

Step 1: For route *j* of the schedule:

Run the NS model and produce the inherent delay estimates, i.e. ${}_e d_{ij}^D$ / ${}_e d_{ij}^A$ for each flight (f_{ij}) in route *j*.

Calculate current schedule reliability, i.e. R_{ij}^D / R_{ij}^A for each flight f_{ij} in route *j* and compare these with the chosen target schedule reliability index, R_{tar}.

Step 2: For each flight f_{ij} in route *j*:

Calculate the initial estimate of schedule adjustment Δs_{ij}^D for f_{ij} by (5.5) and calculate the total available schedule adjustment for route *j*, Δs_j^D by (5.6).

Step 3: Flight re-timing:

If $\Delta s_{ij}^D > 0$ and meets the condition: $(\{d_{ij}^D > 15\} \vee \{d_{ij}^D > {}_{tar}d_{ij}^D\})$, add Δs_{ij}^D minutes (in multiples of five minutes) to the ground time of flight f_{ij}.

If $\Delta s_{ij}^D < 0$, deduct Δs_{ij}^D minutes (in multiples of five minutes) from the ground time of flight f_{ij}.

Step 4: Re-estimate flight delays:

Run the NS model and re-estimate the delays of those flights following f_{ij} on the same route. Re-time $f_{i+1,j}$ and repeat Step 2 and Step 3 for the remaining flights in route *j*.

Repeat Step 1 for each route of the study schedule.

5.5.3 Case Study – Schedule Optimisation by Flight Re-timing

Following the case study airline (Airline P) in the previous section of this chapter, we applied the Sequential Optimisation algorithm to study the potential impact of flight re-timing and schedule optimisation on improving schedule reliability and robustness. The chosen reliability target of schedule optimisation in the following case study was to meet either one of the following two criteria: (1) 70 per cent individual reliability for each flight in the network; or (2) less than 15 minutes mean delays, whichever requires the least buffer time. Schedule adjustments in optimisation were only made to the scheduled ground time of flights. In other words, extra buffer time was embedded in the ground time (turnaround time) of flights in order to control departure delays. When schedule adjustments were required for a flight, a unit of schedule adjustment was 5 minutes. This prevented the final schedule from having unrealistic departure times such as a 9:13 departure. This practice also complied with the current policy of airport slot allocation and co-ordination at those slot-constrained airports (IATA 2004).

The developed Sequential Optimisation algorithm was applied to Airline P's network and the summary results are shown in Figure 5.9. The usage of buffer time for each route was compared with the estimated saving of delays after optimisation. Additional 260 minutes were used to "relax" aircraft routing plans according to optimisation criteria given above, and this generated an estimated delay saving of 540 minutes network wide. Delays of some flights in some of the routes were significantly reduced after optimisation such as aircraft No. 2, 3 and 6, while the impact of minor schedule changes on some routes such as aircraft No. 8 and 11 was less than others. The significant delay reduction effect (being the surrogate of punctuality improvement) after Sequential Optimisation also revealed the extent to which delay propagation has had impacts on current operations by Airline P. A rough estimate of the impact scale of delay propagation from results above was that one minute of original delay may approximately cause two minutes of network-wide delays for Airline P.

Regarding the benefits/costs of applying flight re-timing optimisation, if we assume that the monetary cost of one unit of buffer time and delay time is both $200 dollars per minute, then the estimated impact of the Sequential Optimisation on the new schedule is equivalent to an extra expenditure for Airline P by $19 million dollars per annum (calculated from $200/mins@260mins/day@365days/year) due to lower aircraft productivity. However, the potential delay cost saving could be as high as $39 million dollars per annum (from $200/mins@540mins/day@365days/year), resulting in a $20 million dollars net saving on operating costs per annum as well as a more operationally reliable network. This estimated cost saving could be significant enough for the regional fleets of Airline P and could easily have a positive impact on the profitability and operational reliability of an airline.

The optimised schedule was further simulated by the NS model to evaluate the performance of flight re-timing on improving schedule reliability. Simulation results from the NS model were compared with the current delays of the Reality Case and the inherent delays of the Dream Case in Figure 5.10. We can observe that the optimised schedule effectively controlled the overall delays across the network to the required target level. Total departure delays of the original schedule were 1,816 minutes, which was reduced to 1,278 minutes after optimisation. This result also reflected the increase of the average network-wide schedule reliability from 37 per cent to 52 per cent after optimisation.

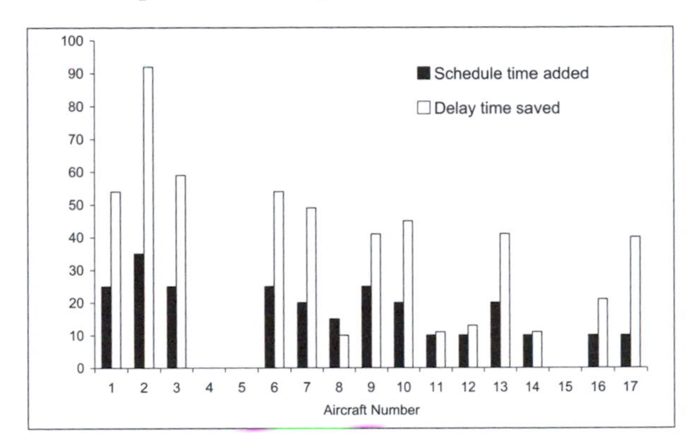

Figure 5.9 Flight re-timing and potential impact on delay saving

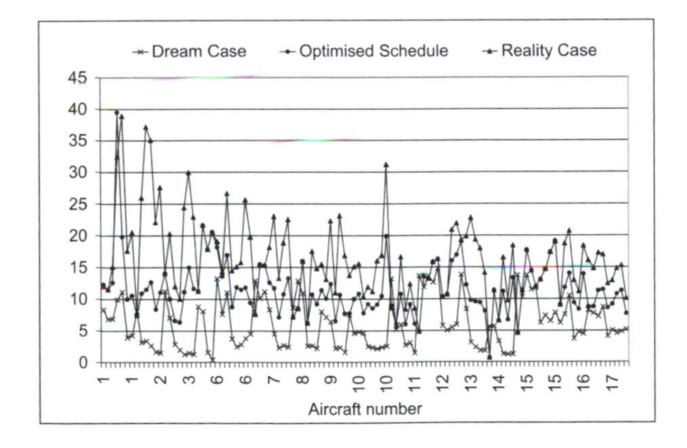

Figure 5.10 Simulation results of the optimised schedule

To demonstrate the effectiveness of Sequential Optimisation on improving schedule reliability and reducing delays, the optimisation results of Aircraft No. 1 were given in Figure 5.11, comparing the reliability index before and after optimisation. We can see that the original reliability of the route decreased from 70 per cent at the start of the cycle to 25 per cent at the end of the cycle. This implies that the actual delay level of this route was higher than the inherent delay level expected from Airline P's schedule. Two possible explanations are: either the inherent delay level was too low because Airline P was too optimistic on the operation of this route; or, the inherent delay level reflected the expectation, but the actual operations of these flights were not reliable enough or were subject to frequent disruptions and delays.

When flight delays before optimisation were compared with those after optimisation for this study route in Figure 5.12, it can be seen that delays after optimisation were better controlled within the range between 15 and 20 minutes. Schedule adjustments to this routing cycle included: five minutes to Flight 103, 15 minutes to Flight 104 and 5 minutes to Flight 105, totalling extra buffer time of 25 minutes for the route by Aircraft No. 1. The usage of buffer time in optimisation reflected closely the current operating status of study flights as shown in Figure 5.12, so more buffer time was deployed for those flights that suffered higher delays than for the others. The estimated average delay time saving after optimisation was 54 minutes for this route, which brought net positive benefits (in terms of delay cost saving) to Airline P.

The major advantage of the proposed Sequential Optimisation algorithm was to utilise the delay/punctuality propagation in a closed network so as to utilise limited schedule time and prevent the use of excessive buffer time during the optimisation process. To demonstrate how punctuality propagation can help Sequential Optimisation and illustrate the convergence of the proposed optimisation algorithm, interim results of the optimisation of Aircraft No. 1 were extracted and

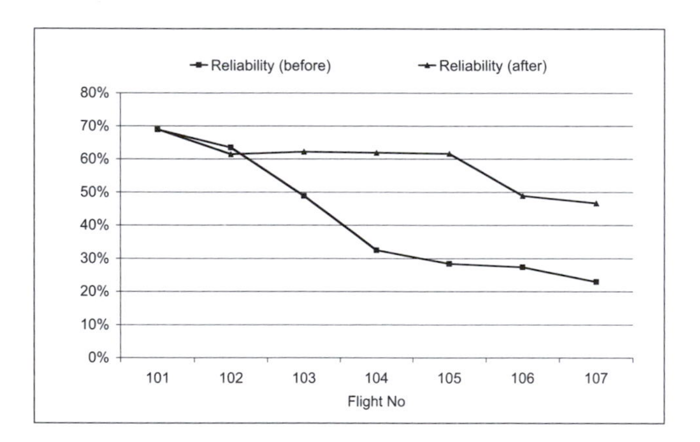

Figure 5.11 Schedule reliability before and after flight re-timing optimisation

shown in Figure 5.13. Current operating delays in this route increased significantly after Flight 104, so subsequent flights in the route suffered from severe delay propagation. The "target delay level" was calculated based on the target reliability (70 per cent) and the inherent delays of this route that were strongly influenced by the current ground operations and scheduling policy of Airline P.

Flight re-timing optimisation started with Flight 101. The initial estimate of the total schedule adjustment time for all flights in this route was 66 minutes. The optimisation procedure first made a 5-minute change to Flight 103. From Figure 5.13 we can see the simulated delay of the route after the first schedule change was reduced from the current level and consequently the re-calculated total schedule adjustment for following flights dropped from 66 minutes to 50 minutes, due to "punctuality propagation". After adding 15 minutes to Flight 104, the delay level of the route dropped further, causing the required total schedule adjustment time for following flights to be only 17 minutes. The impact of punctuality propagation in this route was best seen by Flight 104, and we can also observe how significantly delays of the flights following after 104 were reduced.

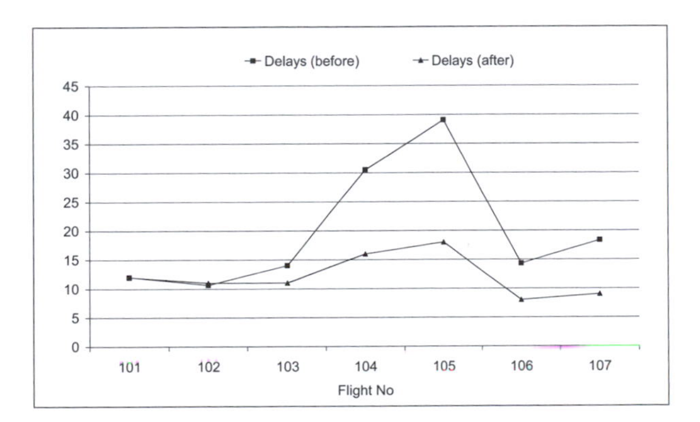

Figure 5.12 Flight delays before and after flight re-timing (Aircraft No. 1)

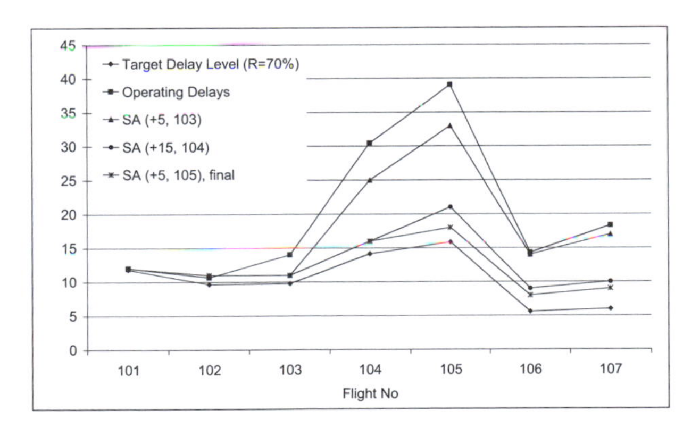

Figure 5.13 Interim results of flight re-timing optimisation (Aircraft No. 1)

The optimisation process stopped after adding 5 minutes to Flight 105. At the end, the added schedule time in the optimisation was 25 minutes, which was significantly lower than the first estimation in the beginning of optimisation, i.e. 66 minutes. Without the Sequential Optimisation algorithm, the application of the initial 66 minutes to the route would certainly improve the reliability of each of the flights more significantly, but would result in spending an extra 41 minutes when compared with the results of Sequential Optimisation. A similar study by United Airlines (UA) developed an algorithm to allocate different ground time for aircraft turnarounds according to the delay histories of individual flights at different airports (Sandhu 2007). That algorithm was essentially similar to the Sequential Optimisation algorithm we have developed here. However, without considering delay/punctuality propagation in the model, the UA algorithm would have over-compensated for delays and eventually deployed too much buffer time than required. By the Sequential Optimisation algorithm, airlines can optimally utilise valuable aircraft time on critical flights in the network to effectively achieve the target schedule reliability.

In practice, the immediate challenge before finalising aircraft routing plans is to match scheduled departure/arrival time of flights with the availability of airport slots. For flights departing/arriving at those airports served by an airline with high frequency, the airport slot availability issue is a relatively minor concern as the airline can always swap slots among its own flights. Since the magnitude of flight time adjustment in the above optimisation is generally below 15 minutes, this change would not impose too much pressure on airport slot availability, even though for an airline that does not enjoy a dominant position at some airports.

The further implication of airport slots is the "saleability" of flights to passengers. Since the Sequential Optimisation or any manual changes during aircraft routing planning and schedule generation always involves altering the schedule, the optimised schedule is not necessarily the most saleable one for potential passengers. There is a possibility that marketing/sales decisions may overturn some results of aircraft routing with specified departure/arrival time requests for certain flights. This practice is quite often seen in the airline industry, especially for airlines operating hubbing schedules due to the need to consider flight connection times and connecting opportunities for passengers at hub airports. The present Sequential Optimisation algorithm has not yet fully considered the constraints airlines may face when dealing with hubbing and schedule synchronisation across the network. Due to this limitation, the applicability of the current algorithm seems more beneficial to airlines which do not operate highly synchronised hubbing schedules. However, the proposed algorithm can be generalised to consider hubbing schedules in future work by using other side constraints in optimisation, if needed.

Considering the operation of strong hubbing activities, there has been growing concerns over the economic benefits/costs for strong hubbing schedules and the congestion costs imposed on airports and air passengers (Goolsbee 2005). The de-peaking of hubbing schedules recently by American Airlines at Chicago O'Haire and Dallas-Fort Worth (DFW), by Lufthansa at Frankfurt, and by Delta Air Lines at Atlanta, demonstrated the vulnerability of heavy hubbing operations in terms of operational reliability of airline schedules and the hidden costs to operate such a network (Field 2005; Goedeking and Sala 2003). The trend in airline scheduling has gradually shifted from maximising connecting opportunities and network traffic in the past, to maximising the operational reliability of airline schedules under the influence of disruptions, while key flight connections are still preserved (Kang 2004; Lan et al. 2006). Hence, airlines are now more willing to trade off hubbing and aircraft utilisation for reliable and robust hubbing (in a weak hubbing or rolling hub form), or more point-to-point schedule operations (Mederer and Frank 2002; Wu 2005). We shall discuss the emerging trends of airline scheduling in more detail in the next chapter.

5.6 Summary

This chapter has discussed airline operations from both strategic and tactical perspectives. From a tactical perspective, this chapter has focused on two major operational issues, namely schedule disruption management and schedule on-time performance (OTP) management. On the disruption management side, we have discussed the causes of operating disruptions in day-to-day airline operations, the consequent scales of disruptions and their impact on airline operations. Commonly used disruption management tactics by airlines were introduced as well as the pros and cons of individual disruption management tactics when deployed in actual situations.

On the OTP management, we have discussed some widely used OTP measurement statistics. The legal background of OTP statistics reporting was discussed for three cases, including Australia, the U.S. and the European Union. A case study on OTP analysis was provided to demonstrate how OTP analysis could be approached, depending on the availability of airline data and the objectives of OTP analysis. Based on the delay code analysis of the study case, we localised potential root causes of delays and their impact on the network operation of Airline Z. The reasoning used in this case study on Airline Z can be easily generalised by airlines and applied to other networks to improve schedule reliability and robustness in future operations.

In the last two sections of this chapter, the concept of *inherent delays* of an airline schedule was defined and further explored by using an actual case, Airline P. The case study on Airline P's network demonstrated how the "inherent delay" concept could be utilised by airline schedulers to investigate the gap between actual operating results (measured by delays and OTP) and the expected operating outcomes. Following the inherent delay concept, a flight re-timing optimisation algorithm, namely the *Sequential Optimisation* algorithm, was developed. This algorithm provided airlines with a tool to adjust flight time according to past operating OTP history and the expected delay level from schedule planning. A case study on Airline P's network demonstrated the effectiveness of the flight re-timing optimisation model and the potential cost savings it could bring to an airline.

In the next chapter, we will focus on discussing the emerging trend of airline scheduling, namely *Robust Airline Scheduling*. This scheduling concept combines airline operations within airline scheduling, like the concept of inherent delays we have introduced in this chapter. We will examine the emergence of this new scheduling concept, the background of the emergence and the state-of-the-art application of this scheduling philosophy in actual airline schedule optimisation. Before finishing Chapter 6, we will provide readers with some ideas for deploying these new scheduling concepts and how this may change the way airlines design and operate networks in the future.

Appendix

Notations and Symbols Introduced in Chapter 5

R_{ij}^{D}	denotes the departure reliability of flight f_{ij}
R_{ij}^{A}	denotes the arrival reliability of flight f_{ij}
R_{ij}	the overall operational reliability of flight f_{ij}
R_{j}	represents the reliability of route j
R_{net}	represents the network-wide reliability of a schedule
R_{tar}	the target schedule reliability
${}_{e}d_{ij}^{D}$	represents the expected departure delay of f_{ij} in the Dream Case
${}_{e}d_{ij}^{A}$	represents the expected arrival delay of f_{ij} in the Dream Case
J	the set of routes in an airline schedule
F	the set of flights in an airline schedule
${}_{tar}d_{ij}^{D}$	the corresponding target departure delay based on R_{tar}
Δs_{ij}^{D}	the amount of flight re-timing for f_{ij}
Δs_{j}^{D}	the total available schedule adjustment time for route j
$\left(d_{i+1,j}^{D}\right)'$	the revised expected departure delay of $f_{i+1,j}$

Chapter 6

Robust Airline Scheduling and Operational Reliability

The focus of this chapter is the emerging concept of airline scheduling, namely *Robust Airline Scheduling*. This chapter starts by firstly providing a brief review of past practices of scheduling in the airline industry. Points are raised as to how past practices have evolved and developed, and the weakness of this conventional paradigm in actual airline operations. Based on the review, this chapter will then introduce the concept of schedule robustness in the context of airline business in Section 6.2. Drawing from experiences of other industries, this concept and its evolution are examined from the perspectives of airline scheduling and operations. Both strategic and tactical improvements are discussed in Sections 6.3 and 6.4, in light of improving schedule robustness in airline operations. Recent advances in robust airline scheduling and integrated modelling are examined for improving strategic airline scheduling. Tactical options to stabilise airline operations in a random environment are also outlined in Section 6.4.

Before finishing this chapter and concluding this book, we return to the debate of airline schedule planning and its on-going struggle with the complexity of network synchronisation as well as with the market-driven economic forces in corporate finance. Discussions of the root causes of this struggle are focused on the dilemma that many airlines face: balancing the gains from network economics, e.g. corporate revenues, and the potential losses that come with these gains due to delays and disruptions. Also examined is the success of the low-cost airline business model with its simplicity in design and operations, and its long-term sustainability in fluctuating market situations. Before concluding this book, we examine the prevailing trend of the schedule planning paradigm in the industry and explore some aspects of the potential future scheduling philosophy.

6.1 The Conventional Paradigm of Airline Schedule Planning

Airline schedule planning as discussed earlier in this book is often conducted by four sequential (and sometimes iterative) stages, namely schedule generation, fleet assignment, aircraft routing and crew pairing/rostering. This conventional paradigm of schedule planning has been mostly driven by and formed due to the fact that the entire airline schedule planning task involves too many challenges such as the complexity of resource synchronisation between the multiple "layers" of the network, and numerous constraints on aircraft deployment and crew rostering.

Since it is mathematically difficult, although not impossible, to solve the entire airline schedule planning problem in a single model, the conventional approach by airlines has been to "divide and conquer": the entire scheduling task is divided into four major tasks and these are solved separately and sequentially.

Schedule generation is conducted first in order to identify and forecast market demands in the near future. Route planning (or network planning) is an integral part of schedule generation, so some new markets are added to the network and flight frequencies to various markets are adjusted according to market updates. The outputs of schedule generation (a preliminary timetable) are used as the inputs to the next stage, fleet assignment. Since different aircraft fleets have different levels of operating costs and seat capacity, the fleet assignment task is to solve the minimum-cost assignment problem in order to assign the most suitable type of aircraft for individual flights in the timetable while meeting the maintenance requirement of aircraft.

Results of fleet assignment are then used in aircraft routing optimisation. This task is to utilise available aircraft in each fleet to carry out flights at the right time and between the right airports. The outputs of aircraft routing are the assignment of flights to individual aircraft, forming "routes" for each aircraft, so flights on the same route are conducted by the same aircraft in a chronological order, according to the schedule. Given the constraints of piloting an aircraft, crew need to be "paired" with different types of aircraft; hence, the crew pairing and the crew rostering problem. Following the results of aircraft routing, crew are paired accordingly with aircraft, but are subject to more strict pairing conditions due to working rules and safety considerations.

The sequential nature of these four tasks also reflects the general timeline of airline scheduling. Schedule generation starts one season (about six months) before the publication of a new timetable for the new season. Fleet assignment starts three to four months before the scheduled flight time and this result also provides the revenue management system of an airline with the information of seat supplies to facilitate inventory and revenue management. Aircraft routing starts approximately one to two months before the scheduled flight time; crew pairing and crew rostering starts around the similar time, and is often done somewhat in parallel, due to the close reliance of crew pairing on aircraft routing results. Crew rosters are often published one roster period before execution; a roster period can be 28 or 30 days, depending on the policy of an airline.

As we will see in the following sections, each of the four stages is a mathematically challenging problem. Given the timeline of airline schedule planning, it is quite reasonable that the entire scheduling problem is divided and solved sequentially and iteratively. However, there are some shortcomings with this conventional scheduling paradigm as well. We will discuss this paradigm in detail as well as its shortcomings in the following sections. The paper by Barnhart et al. (2003) provides a good discussion and overview on the application of operations research (OR) in the airline industry.

1.1.1 Schedule Generation and Fleet Assignment

Most literature on airline network design and schedule generation focuses on the shapes (or types) of networks, the flight frequency planning issue and the fleet choice problem (see for example: Erdmann et al. 2001; Ghobrial and Kanafani 1995; Hsu and Wen 2000; 2003; Lederer and Nambimadom 1998; Teodorovic 1983; Teodorovic and Krcmar-Kozic 1989; Teodorovic et al. 1994; Trietsch 1993). Typically in these works, the uncertainties in airline operations are often not considered or hardly enter into the schedule design models. From the perspective of schedule planning, this approach is fully justified because the issues of most concern at this planning stage are often demand forecasting, flight frequency optimisation, and fleet choices (i.e. aircraft capacity).

The fleet assignment model (FAM) has received high attention in the industry, because FAM essentially determines how demands of individual flights (sometimes also "through-flights") are met by available fleets, and this largely determines how profitable an airline timetable will be after operations. Classic FAMs such as those produced by Abara (1989) and Hane et al. (1995) focused on the mathematical formulation of the fleet assignment problem as well as on the development of solution algorithms. FAM has been mathematically challenging for airlines for two reasons: first, the potential revenue of flights are uncertain, although the cost of operating aircraft is largely certain (but still difficult to precisely calculate); second, the integer programming problem of FAM is often too large to solve without developing ad-hoc solution algorithms (Klabjan 2005; Klabjan et al. 2001; Sherali et al., 2006).

FAM has evolved from considering individual flights, to considering "through flights", and further to a basis of considering passenger itineraries. A through flight is a flight which is conducted by one aircraft but involves two segments in the whole journey. It means that this flight starts at Airport A and ends at Airport B via Airport C, so as to pick up the traffic from C to B, while maintaining the "through traffic" from A to B aboard at Airport C during turnaround. The capture of through traffic can be significant in some networks. The recent consideration of passenger itineraries has introduced a new dimension to FAM, which reflects more significant passenger itinerary influences and network effects on FAM results (see: Barnhart et al. 2002; Barnhart et al. 2009).

1.1.2 Aircraft Routing and Crew Pairing

The aircraft routing problem has been solved as a network flow (or multi-commodity flow) problem in the past, see for example: Cordeau et al. (2001) and Desaulniers et al. (1997), or a travelling salesman problem with side constraints such as Clarke et al. (1997). The challenges embedded in aircraft routing in these models were complex constraints and a large problem size, which were hard to solve. More recently, aircraft routing problem has been solved as a "set-partitioning" problem. Instead of having numerous constraints, the "string" model represents each feasible routing (feasible for maintenance) as a "string" in the model formulation and goes on to solve the aircraft routing as a set-partitioning problem (Barnhart et al. 1998). This type of string model has numerous decision variables, so algorithms such as the column generation method are widely used to efficiently solve string models in aircraft routing. An advantage of a string model is its potential to integrate aircraft routing with FAM such as Barnhart et al. (1998). We will return to discuss integrated models later in this chapter.

Crew pairing has been well studied in the past and a good review is provided by Barnhart et al. (2003). Crew pairing is inherently difficult for two reasons: first, there are numerous crewing requirements and restrictions to meet, so the mathematical problem itself is large and difficult to solve; second, crew pairing has a close dependence on aircraft routing, so traditionally the solution quality of aircraft routing also influences that of crew pairing. In the past, crew pairings were produced by pairing generators. Due to the nature of such a combinatorial problem, there were numerous possible pairings, so these pairings were "filtered" according to scheduling policies, feasibility and legality of crewing (Klabjan et al. 2001). Recently, crew pairing has been modelled as a "set-partitioning problem" which is aimed at determining a set of "pairings" (also known as a tour of duties or crew itineraries) that are able to cover all flights with the minimum crew costs and meanwhile, meet crewing requirements and constraints (Marki and Klabjan 2004; Ryan 1992; Vance et al. 1997). To improve the solution quality and save crew costs, the consideration of aircraft maintenance routing was brought into the modelling of crew pairing (Cohn and Barnhart 2003).

1.1.3 Operational Issues Borne with this Paradigm

The advances in mathematical computing and the progress in the Operations Research (OR) field of mathematics over the past decades have solved many airline scheduling problems to optimality that were unsolvable before. However, experiences have shown that many operating issues in airline business are in actuality caused by this scheduling paradigm. The first weakness of this scheduling paradigm is that it is lacking in consideration of the stochastic effects in a complex airline network. As discussed earlier in this book, stochastic forces play a key role on influencing daily airline operations. Stochastic disrupting events can occur to any elements of an airline network, from aircraft operations, crewing, to passenger handling. Although disruptions can be minor or major in terms of the duration of a disruption, the network effect of a complex network implies that any event or a collection of events may propagate across the system and significantly affect the performance of the system as a whole. Delay propagation is one of these network effects.

The four-stage approach commonly used in airline scheduling does not consider the influence of stochastic forces in operating a complex system. The conventional approach solves the optimisation tasks very well and the optimisation problem at each stage can be solved to nearly the optimum. However, the result of this practice is to create a compact network with a high degree of synchronisation between sub-networks such as the one between aircraft routing and crew pairing. This means that such a network runs well in ordinary operating conditions, but may fall apart amid disruptions, unless significant efforts are made to deal with disruption management and re-scheduling. A good survey of disruption management and irregular airline operations can be found from Ball et al. (2007). Limited consideration of stochastic events may be included in this paradigm and the use of schedule buffer time as the dampening element in a complex airline network has been the common practice in recent projects (Wu 2005).

The second weakness of this conventional approach is the low planned cost of an airline schedule, with high operating cost after execution. Since every stage of the approach is dealt with as an optimisation problem, the result of airline scheduling implies that the planned cost of this system is at a minimum, considering those restrictions imposed during the optimisation process. As mentioned earlier, such a highly synchronised network can be disrupted and the consequences can be severe in some cases. In such a scenario, when irregularities do occur, the airlines must provide extra resources to recover the schedule, resulting in extra operating expenses. Hence, the more often disruption management is required, the higher the operating cost will be for an airline. Although this paradigm can produce airline schedules with a low planned cost, in many cases the operating cost of such a system is high.

The third weakness of this approach is the solution quality of optimisation. Due to the sequential nature of the four-stage approach, airline schedule planning has been done in such a way that solutions of a previous stage constrain the potential solutions of the following stages. Moreover, the solution quality of a stage is highly dependant on the model itself as well as the constraints that are considered in the model. Hence, different demand forecasts will affect schedule establishment. This forecast of scheduling demand influences the result of fleet assignment and the profitability of such a plan. FAM outputs may change aircraft routing and subsequently impact crew pairing solutions. In some cases, a feasible aircraft routing solution cannot be obtained by directly using the outputs of FAM, because the solution "space" available for aircraft routing has been constrained too much by the outputs of FAM (Barnhart et al. 1998). For the entire airline scheduling problem, such a divide-and-conquer approach usually produces sub-optimal results when the whole system is considered, rather than the solutions for each stage.

6.2 The Emergence of Robust Airline Scheduling

Given the weaknesses of the four-stage paradigm, a new scheduling paradigm, namely *Robust Airline Scheduling,* has emerged that is aimed at addressing the potential impact of uncertainties which occur between schedule planning and actual operations. In addition, an emerging trend, namely *Integrated Scheduling* that aims at improving the quality of airline schedule optimisations will also be discussed in the following sections.

6.2.1 Robust Airline Scheduling – Definition in the Airline Context

There are various definitions for *schedule robustness* in the literature of airline schedule planning and operations research. In general, schedule robustness has two broad definitions according to the objectives of robust scheduling: first, *robust scheduling is to plan against major disruptions*; and second, *robust scheduling is to plan for higher schedule flexibility and reliability against minor disruptions*. Theoretically speaking, the concept of airline robust scheduling comes from the field of *robust optimisation* which is a branch of *stochastic optimisation* in mathematical programming dating back to the 1950s from the classic work by Dantzig (1955). Stochastic optimisation deals with optimisation problems whose parameters follow certain probability distributions or are assumed to take a probabilistic description (Birge and Louveaux 1997).

Robust optimisation, on the other hand, usually deals with a deterministic model and considers a set of scenarios that may rise under uncertainties in actual operations. In the context of airline schedule planning, robust optimisation aims to address the potential issues caused by data uncertainty in a strategic planning model. As was pointed out by Ben-Tal and Nemirovski (2000) in a recent study, optimisation results can be remarkably sensitive to the perturbations in the parameters of the problem; this is also the case for airline schedule planning.

As mentioned earlier, airline operations are subject to stochastic influences from the operating environment. These stochastic forces appear in two forms: uncertain passenger demand and stochastic disrupting events in airline operations. Uncertain passenger demand affects schedule generation and establishment, and subsequently impacts the model results of FAM (hence, profitability), aircraft routing and crew pairing. Operational uncertainty does not influence schedule establishment and FAM, but may affect how aircraft routing and crew pairing are planned and operated. The conventional paradigm of airline scheduling has the weakness of producing a schedule with a low planned cost, but amid stochastic disruptions, the schedule exhibits a high operating cost (in the form of "recovery cost"). Therefore, a definition of robust scheduling is to *consider potential recovery costs from actual operations at the planning stage, so the "realised costs" of schedule planning and schedule execution are minimised* (Ball et al. 2007).

A highly synchronised network often suffers from a weakness of flexibility in operational uncertainties and the conventional airline scheduling paradigm is no different. To address this operational weakness, an emerging definition of robust scheduling is to *design an airline network in which resources connections are well-synchronised and also enjoy a certain degree of "flexibility"* (Lee et al. 2007; Wu 2006). The objective of this robust scheduling is to plan an airline schedule with appropriate buffer space to withstand disruptions (mostly minor ones), so as to involve as few schedule recovery actions as possible in schedule execution. By this definition, the planned cost of a schedule would be higher than that produced by the current paradigm, but would still be lower than the total "realised cost" according to the current sequential approach in scheduling.

6.2.2 Robust Optimisation – Experiences from Other Industries

Robust optimisation has been widely used in other industries. Inventory control is a popular area for application (see for example, Adida and Perakis 2006; Bertsimas and Thiele 2006). In the inventory control problem, demand is uncertain and ordering, stocking, and storage of materials (or goods) can incur extra costs. Hence, the goal is to minimise the total cost under demand uncertainties in such an inventory control problem, as is common to many manufacturing industries, supply chains, and retailers.

Robust optimisation has long been used in the finance sector, in particular on robust portfolio optimisation and modelling market return dynamics. Modern portfolio theory has its roots in the work by Markowitz (1952; 1959), and recent progress in this area such as that by Costa and Paiva (2002) and Tutuncu and Koenig (2004) has made significant contributions to the overall portfolio theory. In the work by Tutuncu and Koenig, they computed "robust efficient frontiers" by using real market data and found that the robust portfolios offered significant improvement in worst-case scenarios compared to nominal portfolios, although there were extra costs in expected return.

What we can learn from the above examples in other industries is that robust optimisation has been proved as a useful tool in dealing with data uncertainties in a planning model. In the airline business context, uncertainties from passenger demand significantly influence airline schedule planning outcomes as well as schedule profitability. Uncertainties in airline operations cause further perturbations to planned schedules. Without considering operational uncertainties in schedule planning, airlines may incur higher operating costs and some disruptions are in fact caused by an airlines' own schedule planning.

6.3 Improving Airline Schedule Robustness: Strategies

Robust scheduling in airline business comes in various forms and addresses different concerns of uncertainties in an airline network, including addressing stochastic demands, incorporating recovery costs in planned costs (for major disruptions), and adjusting schedules for uncertainties (for minor disruptions).

6.3.1 Robust FAM for Stochastic Demand

When addressing the uncertain demand of markets, FAM is an ideal target for improvement, because all subsequent scheduling tasks depend on the results of FAM. Fleet assignment is often done a few months before schedule execution, and the actual market demand is only realised along the timeline of passenger booking until the day of departure of a specific flight. This means that the results of FAM may not be appropriate when the realised demand becomes clear; some flights may have insufficient capacity, and others may have excess capacity, due to optimistic demand forecasts months ago. The operation of such a schedule will result in the loss of potential demand (as revenues) in flights with insufficient capacity, or excessive aircraft operating costs in flights with over optimistic demand forecasts, because the assigned aircraft is too large for the realised demand.

In the past, FAM often used simplified assumptions on passenger demand, revenue calculation and largely overlooked the "network effects" of passenger itineraries. More recently, researchers and airlines have started addressing these weaknesses in FAM. Jacobs et al. (2008) presented a model that was aimed at dealing with network effects and stochastic demands of markets. On improving the past weakness of FAM in this area, Jacobs et al. suggested that FAM should account for multiple markets that utilise each leg of the schedule, multiple classes within each individual market, and network interactions coming from market competition for customer demand. Barnhart et al. (2002) proposed an itinerary-based FAM that was capable of capturing network effects of passenger itineraries, and estimating the spill and recapture of passenger bookings in the network.

The other methodology for addressing demand uncertainty is to use the two-stage (or multi-stage) stochastic programming in solving FAM such as in the work by Sherali and Zhu (2008). The first stage of stochastic FAM is based on demand forecasts and only assigns fleet families to flights. The second stage conducts the family-based type-level assignments according to market demand realisations. In the industry, the aforementioned "second stage" is sometimes done separately from FAM, when booking realisations are available. In such a case, FAM is done as before, while the second stage "recourse" problem is conducted on a selected number of flights on an ad-hoc basis about seven to thirty days before departure. Fleet assignments to those target flights are re-examined and may be swapped with other flights in order to maximise capacity utilisation and also meet potential demands.

Instead of using stochastic programming to revise fleet assignment results, "aircraft swap" is a tactical measure that airlines often employ on an ad-hoc basis to deal with uncertain passenger demand. Similar to the timeline of the two-stage stochastic programming, aircraft swaps are suggested based on realised demand and booking curves in airline revenue management tools. However, the focus of aircraft swap is more on cost-benefit evaluation on airline revenues and airline operations, so the "quality" of swaps and the timing of swaps are critical for success. Swaps done early will cause less disruptions (and hence lowering aircraft operating costs), but the swap may eventually become inferior to the originally planned type of aircraft as the departure time draws near (when less late bookings are realised). Late swaps, on the other hand, cause more disruptions and adjustments to aircraft routing and crewing, but have a lower risk of under- or over-estimation of demands. Bish et al. (2004) conducted a series of tests on swap quality and timing issues. Recently, All Nippon Airways and United Airlines both adopted this approach and reported significant revenue recovery compared to conventional FAM results (Oba 2007; Zhao et al. 2007).

Robust FAM can also be achieved by considering operational uncertainties in the implementation of FAM. Rosenberger et al. (2004) proposed the concept of "hub isolation and short cycles". Hub isolation limits the total hub connectivity, while short cycles refers to the construction of routes with limited flight legs before returning to the hub airport. Short cycles are in particular helpful amid flight cancellations, because the impact of cancelling a flight is limited due to the nature of short cycles. Smith and Johnson (2006) proposed an interesting concept, called "station purity", and incorporated this concept with robust FAM. Station purity limits the types of fleets at spokes to at most one or two fleet types. This allows more opportunities for crew swaps when disruptions occurred. An additional benefit to station purity is that it lowers the cost of inventory on aircraft spare parts at individual airports.

6.3.2 Realised Cost Minimisation

As mentioned earlier, the conventional approach of airline scheduling tends to minimise the *planned costs* of a schedule, mostly due to the lack of consideration of uncertainties in demand and operations. The consequence of this approach is that the *realised costs* after operation can be significant. The emerging concept of robust airline scheduling also seeks to address this issue by incorporating into airline scheduling the operational uncertainties and consequent *future costs* due to schedule recovery, and aims to minimise the realised costs at the stage of schedule planning. We will discuss some recent developments in this field focusing on two areas: first, planning for robust airline operations; and second, planning for robust crewing.

Since there are random forces in the operating environment of an airline, any airline plans are subject to random forces, either in the form of uncertain passenger demands or stochastic disruptions in operations. Rosenberger et al. (2002) developed an airline simulation model, called SIMAIR, to address airline operations in a random environment. In SIMAIR, stochastic factors were modelled together with airline schedules, recovery policies and performance measures. This model was particularly helpful to evaluate the performance of various schedule recovery options and also to estimate the future costs of operational disruptions and recovery actions.

For a robust airline operation, aircraft routing plays a key role. Due to the fact that flights are assigned to form routes for individual aircraft, the execution performance of an individual flight is closely linked with other flights on the same route via aircraft routing. A poorly designed aircraft routing plan will result in disruptions, often in the form of delays, which then propagate along the aircraft routing. This delay propagation phenomenon can further trigger chain reactions such as passenger itinerary disruptions and broken crew connections. There are various ways of dealing with this issue. Kang (2004) proposed the conceptual "degradable network" in which flights were partitioned into various "layers" and flights in different layers were not connected in aircraft routing (although passengers still did), so disruptions could be contained within individual layers. These layers were associated with different levels of recovery priority. To compensate flights on layers with lower recovery priority, tickets were sold at a lower price; similarly, flights with a higher recovery priority were sold to passengers with a premium for service quality assurance, i.e. quick disruption recovery. The underlying robust routing mechanism was to partition flights in aircraft routing in such a way that when disruptions occurred in actual operations, airlines could easily contain the impact to individual layers and also protect those high-yield flights through quick recovery.

A similar approach, by Ageeva (2000), shifted the focus of robust aircraft routing onto creating "aircraft swap opportunities". The concept was that two aircraft routes could be swapped if they crossed each other more than once along the timeline. By maximising aircraft swap opportunities in aircraft routing planning, airlines enjoyed more "flexibility" in dealing with disruptions in actual operations, and there was a lower demand to adjust aircraft routing due to flight delays or cancellations. This concept, by providing more aircraft swap opportunities, had lower costs in actual operations than other recovery options such as re-scheduling routing and crewing.

Regarding planning for robust crewing, Ehrgott and Ryan (2002) suggested that crewing robustness can be improved by providing a trade off between the crew pairing costs (the planned costs) and an extra amount of integrated future costs at the planning stage. They proposed the measure of "non-robustness" which approximated the potential effects of delay propagation through crewing, in particular due to limited crew connection time and the requirement to change aircraft. According to computational results, this approach could significantly reduce the non-robustness of a schedule by two fold, if the planned costs were raised by one per cent. The cost increases came from longer connection times for crew between flights on different aircraft and longer duty times. In a recent project, this concept was further exploited by Weide et al. (2007) to improve the solution quality of crew pairing by integrating this process with aircraft routing planning.

Yen and Birge (2006) approached robust crewing by stochastic programming and with the goal of minimising the *realised costs* of crewing. They considered the conventional crew pairing problem as the first step of stochastic programming and included future unplanned crewing disruptions and costs in the "recourse" problem of the stochastic programming model. Schaefer et al. (2005) considered "crew pushback" and used Monte Carlo simulation techniques to estimate future costs due to disruptions. Like the aircraft swaps concept, Shebalov and Klabjan (2006) suggested that robust crew pairing could be achieved by designing a "move-up crew" condition in crew pairing solutions. A "move-up crew" shared the same crew base with the departure airport of a target flight and must be able to swap with the crew which was originally assigned to operate the target flight. The objective of move-up crew was to design disruption solutions in crew pairing and provide an airline with operating flexibility in dealing with crew recovery.

6.3.3 Schedule Adjustments for Operational Uncertainties

Uncertainties in operations are dealt with by strategic means as well as by tactical means. Operational uncertainties influence airline operations and cause major and minor disruptions. In robust airline scheduling, stochastic major disruptions are considered by incorporating future anticipated recovery costs in the planned cost during strategic planning. Minor disruptions, on the other hand are dealt with by a different approach that aims at producing a delay-tolerant schedule amid the minor disruptions in daily operations. Since airline operations are subject to stochastic forces, the consequence of such an operation is that the on-time performance (OTP) of flights is also stochastic. Airline timetables are fixed after publication and flight times may not be easily adjusted for many reasons such as airport slots and resources allocation. In order to improve operational robustness in the face of minor disruptions, schedule re-timing is a common method airlines adopt as a strategic means to handle operational uncertainties.

The underlying concept of schedule re-timing is to adjust the length of the turnaround time of flights according to delay histories of individual flights as well as the propensity of delay propagation through aircraft routing. Commonly, the schedule re-timing technique is applied after aircraft routing and crew pairing, so the re-timing is limited to schedule adjustments that will not alter the results of aircraft routing and crew pairing. For instance, Lan et al. (2006), Wu (2006), Lee et al. (2007), and AhmadBeygi et al. (2008b) applied schedule re-timing (also called "schedule fine-tuning") to given aircraft routing plans. The subtle difference between these works was regarding how delays and delay propagation were estimated in an optimisation model.

The work by Wu and Lee et al. used simulation models to estimate delays due to disruptions and the subsequent delay propagation in a network. Lan et al. estimated propagated delays on an individual flight basis and AhmadBeygi et al. improved Lan's methodology by using "delay trees" to trace delay propagation. The difference between these two approaches is that an analytical model of delay propagation such as those employed by Lan and AhmadBeygi necessarily assumes that the delay distributions of flights are independently (and sometimes identically) distributed, namely "i.i.d" distributions. This assumption caused both delay over-estimation and under-estimation in different scenarios (AhmadBeygi et al. 2008b).

As pointed out by Wu (2006), the weakness of the other models, although they have already considered delay propagation, is that they fail to address the following two situations: (1) when two flights are operated in the same route for the first time without precedence (a non-i.i.d. situation); and (2) the situation where punctuality may also propagate due to improvements of OTP of "up-stream" flights in a route.

The first situation causes problems in an analytical model, because one needs to assume the delay distribution of inbound flights without any historical delay data. The second situation occurs when one omits the fact that schedule re-timing can improve the OTP of flights and consequently, alter the distributions of delays of following flights, causing the i.i.d. delay assumption to fail for most cases. Simulation models such as those by Wu (2006) and Lee et al. (2007) are superior in this regard.

The common factor among these flight re-timing models was that they sought to produce a schedule that was resilient to delays and minor disruptions and ideally, was able to contain the impact of delay propagation without turning to costly disruption management actions. According to Lan et al. (2006), allowing a 30-minute time window (i.e. flight times can be 15 minutes earlier or later) for flight re-timing could save 20 per cent of expected passenger delays and reduce 40 per cent misconnections of passenger itineraries. Wu (2006) applied flight re-timing on a small network and reported that the schedule reliability index improved from 37 per cent to 52 per cent, with an estimated delay cost saving around $20 million dollars per annum in a case study.

6.4 Integrated Models for Improving Solution Quality

As we have mentioned earlier, the four-stage paradigm uses the outputs of the previous stage as inputs for the following stage, so the solution quality of later stages, e.g. aircraft routing and crew pairing becomes highly dependent on results of earlier stages. Moreover, the optimisation problem itself can become very difficult to solve due to the large set of constraints imposed at each stage, as well as the limits imposed by input parameters from earlier stages. Hence, there is always a search for integrated models that can change the scheduling paradigm and provide further insights on solution quality. Recently, there are emerging efforts towards this goal that combine some stages of airline scheduling. We will discuss some recent progress in this area following the order of the four-stage scheduling paradigm.

6.4.1 Integrating Schedule Generation and FAM

In the four-stage approach, schedule generation plays a key role in setting up the "scope" of airline scheduling in the form of constraining flight times and frequencies (i.e. demands). This stage is critical in that it influences how the following three planning tasks are done, as well as the optimisation quality of these tasks. Recognising the importance of allowing flexibility in schedule generation, Levin (1971) proposed the idea of adding "time windows" to flights so as to allow flexibility at later planning stages such as aircraft routing. Following this, a number of authors have recently applied this concept (a.k.a. flight re-timing) at various stages of airline scheduling with the same goal: relaxing the constraints of schedule generation and expanding the scope of schedules.

Desaulniers et al. (1997) used time windows for flight departure time in FAM which was formulated as a multi-commodity flow problem with extra time variables. Rexing et al. (2000) employed the time window concept in FAM and simultaneously addressed the assignment problem of fleeting and the setting of flight departure times. Conversely, Klabjan et al. (2002), used the time window concept in addressing airline crew scheduling problems. The time window concept in this case provided much flexibility which is often needed for the already constrained crew pairing problems.

Lohatepanont and Barnhart (2004) extended the itinerary-based FAM of Barnhart et al. (2002) on schedule generation. The goal was to optimise schedules according to the worth of a particular itinerary by removing certain legs and adjusting demands afterwards. Departure times of flights are flexible in this combined scheduling and FAM, and the service frequency and fleet assignments were also simultaneously optimised. Belanger et al. (2006) further expanded FAM to consider the impact of flight re-timing among flights that served the same markets, avoiding "internal" competition among flights of an airline. Sarmadi (2004) combined flight re-timing with aircraft routing and proposed that by doing so, although the planned costs of the airline schedule may increase, the savings from disruptions far outweighed costs. Sarmadi's model optimised flight times as well as aircraft routing by distributing slack time in the network optimally and reducing passenger delays.

It seems from the literature that modifications of the schedule generation process itself or of flight times has the advantage of relaxing an airline network, so it is better able to withstand minor stochastic forces in operations. The time-window idea is particularly helpful in addressing the issue of flight delays and itinerary-based demand in FAM. The work by Lan et al. (2006) and Wu (2006) discussed earlier in the last Section demonstrated the potential benefits of reducing delays by flight re-timing, while Lohatepanont and Barnhart (2004) have also shown how flexible flight time can help improve FAM results.

6.4.2 Integrating FAM and Aircraft Routing

The integration of FAM and aircraft routing has two advantages. Barnhart et al. (1998), in the pioneer study, showed that FAM could be fully integrated with the aircraft routing model. As discussed earlier, the sequential stages of airline scheduling have a weakness in that the solutions of later stages are limited and are often sub-optimal due to constraints imposed by earlier stages in the process. The first advantage of integrating FAM and aircraft routing is that it is then always feasible to find optimal solutions for aircraft maintenance routing and this is not always true in a sequential planning process.

Rosenberger et al. (2004) extended the work of Barnhart et al. by improving the operating robustness of aircraft routing by means of built-in "short-cycles" in the integrated FAM and aircraft routing solution. While enjoying the benefits of feasible maintenance routing, this extension used short-cycles to enhance the flexibility of disruption management of airlines. Hence, the second advantage of integrating FAM and aircraft routing is that airlines have more options in disruption management by utilising those built-in short-cycles and the extra operating costs due to disruption management are lower than before.

More recently, Haouari et al. (2009) proposed the use of a network flow-based model to approach the integrated problem of FAM and routing. By using network flow-based heuristic approaches, Haouari et al. reported that they were able to yield near-optimal solutions with significantly less computing burden. A possible extension on this network flow-based model is to further integrate the concept of "itinerary-based FAM", as proposed by Barnhart et al. (2002). Given the similar network structure, it is possible to apply the itinerary-based FAM and integrate this with aircraft routing. This approach will achieve better results for FAM and jointly for the integrated problem, because the itinerary FAM is capable of reflecting the nature of passenger demand that is based on individual itineraries between markets.

6.4.3 Integrating FAM and Crew Pairing

Since crew pairing is the last stage in the scheduling paradigm, it "inherits" constraints of earlier stages by taking outputs of earlier stages as model inputs. This limits the options one can have in crew pairing, in addition to the fact that the crew pairing problem itself is inherently hard to solve due to the size of most industrial problems and the complex crewing constraints commonly seen in the industry. There are relatively few works that directly integrates FAM and crewing. Sandhu and Klabjan (2007) developed an integrated model for FAM and crew pairing, and meanwhile used "plane-count" constraints to ensure aircraft routing feasibility. They solved the crew pairing problem first and identified those crew connections that required crew to stay on the same aircraft, namely "forced turns". Those forced turns were extended into plane-count feasible rotations, if the total number of aircraft on the ground at the time imposed by forced turns did not exceed the plane-count constraints.

Gao et al. (2009) extended the fleet purity and station purity concept by Smith and Johnson (2006) to improve crew scheduling by incorporating crew connection constraints in FAM for integrated modelling and station purity for robustness improvement. This approach did not truly solve FAM and crew paring simultaneously, but rather crew connections were included in FAM to achieve integration. The advantage of using station purity and fleet purity was that it was easier to plan "move-up crew" in crew scheduling as suggested by Shebalov and Klabjan (2006), so more crew swaps could be utilised during disruption management and extra operating cost due to crewing disruptions could be reduced.

6.4.4 Integrating Aircraft Routing and Crew Pairing

To improve crew pairing quality, a number of works have focused on integrating aircraft routing with crew pairing. Cordeau et al. (2001) have shown that short-connection times can have a significant impact on crew scheduling in actual operations. Cordeau et al. (2001) proposed an integrated approach by linking maintenance routing and crew pairing models with a set of additional constraints and solved the problem by Benders decomposition method, in which aircraft routing was solved in the master problem and crew pairing was solved in the sub-problem. By enforcing the condition that the crew were to stay on the same aircraft as in the routing problem, the crew pairing solution could be more robust in operations.

Klabjan et al. (2002) solved the crew scheduling problem with additional "plane-count constraints" from aircraft routing in a sequential process. They also added flight time windows in the optimisation problem in order to allow more flexibility in crew pairing and aircraft routing. Cohn and Barnhart (2003) extended the concept of Cordeau by incorporating key maintenance routing decisions within a crew scheduling problem and developed an extended crew pairing model that integrated aircraft routing and crew pairing. This model did not truly solve the two problems simultaneously, but rather it incorporated a collection of variables in a crew pairing problem that represented the complete solutions to the maintenance routing problem. For instance, when solving the integrated model, if a short-connection pairing is being selected, the model will ensure that the same connection is also valid in aircraft routing.

Mercier et al. (2005) improved the model of Cordeau et al. (2001) by allowing "restricted connections", where crew connect time was longer than the minimum sit-time, but shorter than a given threshold. Whenever restricted connections were selected, both in aircraft routing and crew pairing, a penalty was applied, i.e. discouraging such a selection in optimisation. Interestingly, Mercier's model improved the speed of convergence by reversing the order that the problems were solved: crew pairing was solved in the master problem and aircraft routing in the sub-problem.

Recognising the dependence between aircraft routing and crew pairing, Weide et al. (2007) suggested an iterative method to solve both problems, as well as to improve the operational robustness of airline schedules. They started from the minimal cost solution of crew pairing without considering any routing restrictions (i.e. not taking routing results as model inputs), and produced a larger set of feasible pairings. This result also demonstrated how constrained the crew pairing problem could be when applying the results of aircraft routing as inputs for a crew pairing problem. In order to improve the operational robustness of crew schedule, the connection time between fights in each pairing in Weide's model was maintained with a 30-minute buffer higher than the minimum sit-time required. Any connections that had a sit-time buffer less than 30 minutes were classified as "restricted connections". In the iterative process, aircraft routing was forced to contain as many of those restricted connections as possible, i.e. enforcing the aircraft to "follow" crew wherever possible. With the incorporated buffer time for crewing in the integrated model, Weide et al. (2007) have showed that schedule robustness could be traded off with the planned costs in strategic planning, so higher operating robustness could be gained by a higher planned cost in crew pairing.

6.4.5 Integrating Three Stages

The potential of integrating more stages in airline scheduling is to reduce the planned costs of scheduling by improving the solution quality. Until now, there has been no model that integrates four stages of airline scheduling, as well as incorporating operational robustness in planning. However, there has been significant progress made recently on integrating three stages of scheduling, or combining two stages while incorporating robustness in planning. Mercier and Soumis (2007) extended the integrated aircraft routing and crew pairing model by Mercier et al. (2005), by incorporating flight re-timing options for each flight in a schedule. By relaxing the constraint of flight times by allowing five minutes of forward/backward re-timing, the number of potential flight connections, including short connections, increased 58.4 per cent on average for the seven test schedules, relative to the originally fixed ones. This increase in feasible flight connections yielded significant cost savings in the integrated routing and crewing model in Mercier's model.

Sandhu and Klabjan (2007) integrated FAM and crew pairing, meanwhile incorporating "plane-count" constraints to ensure maintenance routing feasibility. In a recent paper by Papadakos (2009), however, it was showed that the use of plane-count constraints in the integrated FAM and crewing model did not always guarantee a feasible routing solution. Instead, Papadakos introduced stronger FAM constraints to the integrated model and incorporated enhanced plane-count constraints for aircraft routing. Compared with the original plane-count constraints by Sandhu and Klabjan, this enhanced plane-count technique provided stronger routing constraints, resulting in a fully integrated FAM, aircraft routing and crew pairing model. The financial advantage of Papadakos's integrated model was impressive: on seven tested schedule instances, the cost savings compared with semi-integrated models varied from 0.5 per cent to 2.6 per cent, translating to about $3.5 to $24 million dollars saving per year. Although operational uncertainties were not considered in Papadakos's model, the potential of robustness improvement and operating cost saving by incorporating uncertainties can be expected in future integrated models.

6.4.6 Improving Airline Schedule Robustness: Tactics

Corresponding to the two definitions of schedule robustness, there are tactics that airlines adopt to improve the operational robustness of airline schedules. During airline schedule planning, if a schedule is planned for major disruptions, the robustness gains come from those embedded disruption management options. Since resources are highly synchronised in the sequential planning paradigm, any disruption management actions are costly to an airline as well as to passengers (See Ball et al. (2007) for a review). If robustness is designed into a schedule, there exist some planned options with significantly lower costs that can be utilised to handle irregular airline operations. As mentioned earlier, the degradable network design (Kang 2004) is ideal for disruption planning. By protecting high yield flights and recover these flights first when disruptions strike, an airline would be able to reduce recovery costs due to irregularities, and meanwhile protect some revenue via priority recovery. Similar tactics such as limiting hub connectivity, planning short cycles (Rosenberger et al. 2004), and designing aircraft swaps in aircraft routing (Ageeva 2000), have similar effects on improving operational robustness. The cost of crew disruption is a significant burden for an airline. Hence, concepts like station purity (Smith and Johnson 2006) and move-up crew (Shebalov and Klabjan 2006) provide effective options for crew disruption management.

When planning for schedule robustness against minor disruptions and delay propagation, flight re-timing is an effective tool in strategic airline scheduling. Methodologies proposed by various researchers such as Lan et al. (2006), Wu (2006), Lee et al. (2007), and AhmadBeygi et al. (2008b), are effective tools to improve the operational robustness and reliability in actual operations. Tactically, airlines often turn to improving ground handling efficiency when minor irregularities occur. Since most disruptions are minor in scale in daily operations, they are able to collectively cause network instability due to resources synchronisation in an airline network, i.e. through delay propagation. Disruption management tactics, e.g. flight cancellation or re-scheduling are too costly for such cases which occur very often in daily airline operations. An effective tactic to stabilise network reliability in a random and complex network is to stabilise some key stochastic elements in the system such as those in airline ground operations.

The rationale of this tactic is that delays can be "naturally" contained by the designed schedule buffer time in the network, together with the help of improved ground handling efficiency. Theoretically, some delays can be contained by the planned schedule buffer, but the stochastic nature of a complex system implies that the buffer may not be sufficient to contain delays for some serious cases, or further disruptions are caused due to propagated delays from earlier flights. In such a random environment, the strategy to control and stabilise a complex system is to maintain the proper function of key elements. In the context of airline operations, these key elements are those activities that airlines are able to control and influence such as airline ground operations and airline strategic scheduling. Hence, it is of paramount importance that an airline is able to fully control ground operations in a stochastic operating environment. As such, airlines often focus on ground operations with two specific approaches.

First, the turnaround time of flights can be reduced by allocating more ground resources to make up delays. For instance, connecting passengers can be escorted to ensure in-time connection to outbound flights; more ground staff can be allocated to a specific flight in order to "speed up" aircraft turnaround operations, thus creating new buffer time to ease delay propagation onto other flights in the network. Since the cost of most resources in ground operations is largely fixed, re-deploying or concentrating some resources for specific flights on an ad hoc basis would not incur much extra cost.

Second, some turnaround tasks can be "skipped" without significant losses. For instance, belly cargo can be loaded onto the next flight to the same destination, if the cargos are not express ones. Connecting bags can be skipped for a quick turnaround to reduce delay propagation, because the potential saving network-wide from delay propagation and recovery is far greater than the cost of delivering late bags to passengers (although there are inconveniences involved for some passengers). However, there may be concerns about separating passengers and their bags. Since this decision is always ad hoc and there is rarely a chance to plan ahead for such an ad hoc tactical decision in a random environment, the chance of having dangerous goods or explosives in connecting bags transported on later flights is extremely low. This is why airlines do not allow checked-in baggage to travel without its owner aboard. In fact, late connecting bags occur quite often in everyday airline operations, especially for connecting journeys.

6.5 The Future of Airline Schedule Planning

The evolution of airline networks has occurred gradually in the past few decades with two major stimulating events, namely the air transport deregulation in the U.S. and the 9/11 terror attack in 2001. Deregulation stimulated the evolution of some airline networks towards a hub-and-spoke system, in favour of the economical gains brought by the system. The 9/11 terror attack affected the airline industry heavily with a dramatic drop in aviation demand, forcing many airlines to rethink their cost structures and undergo heavy cost-cutting projects (Dennis 2005; Doganis 2005). It was due to a series of cost-cutting projects in the industry after the 9/11 attack that many airlines came to re-examine the financial bottom line and the balance between gains from network economies and the losses and costs due to the very same network structure.

6.5.1 The On-going Struggle of Airline Scheduling

In a hubbing network, an airline relies heavily on the gains from the economies of density and network efficiency to collect and re-distribute passengers in a network. However, the complex design of such a system also brings airlines the associated negative impacts on profitability. The delicate synchronisation among layers of an airline network makes such a system vulnerable in a random environment; disruptions can easily cause airlines heavy losses. However, without using a hubbing system, an airline may lose a critical competitive edge to other airlines in a network economy, unless the airline changes its business model to run a point-to-point network that focuses on individual markets instead of network gains.

The dilemma of strong network gains and potential heavy losses due to complex networks triggered a recent evolution in airline hubbing activities, namely de-peaking and rolling hubs. American Airlines reduced the intensity of hubbing at Chicago O'Haire and Dallas-Fort Worth. Lufthansa de-peaked its hub at Frankfurt, and Delta Air Lines also reduced hubbing strength at Atlanta (Field 2005; Goedeking and Sala 2003). By reducing the intensity of hubbing traffic, airlines trade off some flight connection opportunities (and the potential gains from them) with gains from reduced negative effects due to inefficient resource utilisation at hub airports and excessive operating costs due to delays. This trade-off has been the on-going struggle for hubbing carriers for years, and it seems now that there is a clear message for most airline boardrooms: *bring back simplicity, operational robustness, and on-going cost cutting to the airline business in order to improve the long-term financial sustainability and profitability*.

6.5.2 Bring Simplicity Back to Airline Scheduling

Does "simplicity" imply that there will be no hubbing networks in the future? Evidence from economic theories clearly shows that this would not be the case: the hub-and-spoke system will co-exist with the point-to-point system in a free market (Alderighi et al. 2005). The term, *simplicity* here refers to *the simplicity in airline network design and the subsequent airline operations*. Simplicity has been one of the major principles that many low-cost carriers adhere to (Gillen and Lall 2004). Simplicity may appear in various forms in an airline business including: reducing resource synchronisation levels, less complex network design, and on-going cost cutting efforts. These various forms of simplicity translate into airline operations by means of lower planned cost in network design, lower operating costs, lower disruption costs, and eventually lead to profitability and sustainability of an airline in the long run.

De-peaking or rolling hubs is a way of bringing simplicity to airline business. By reducing the complexity of resource connections at a hub airport, an airline will focus on maintaining those critical connections (often with high yields) and have to forgo other connections that are not as beneficial. On the positive side, the reduced complexity at a hub simplifies airline operations and reduces the risk and scale of the impact of random disruptions and "self-induced" delays, leading to a lower cost base for operations and profitability. On the negative side, the reduced resource synchronisation means that a number of markets are not served as well as before; passengers may need to wait longer at the hub airport for a connecting flight and the service frequencies of certain markets may decrease accordingly. However, the networks of alliance and regional affiliated partners such as regional feeder services can compensate this potential loss by providing the required network coverage through wet-leases or service outsourcing (Dennis 2005).

The success of some low-cost airlines such as Southwest Airlines and Ryanair is based on the robust and simple airline business model they have adopted. There are three "golden principles" in the low-cost carrier business model: *simplicity, market profitability, and a low cost base*. Simplicity reduces operating costs as well as planned costs for airline resources management. Market profitability drives route planning and the development of a point-to-point network, ensuring that a route is operated if and only if it is profitable, even in the short-run only. Continuous cost cutting ensures that the cost base of the business model does not rise, so lower airfares can be offered to stimulate market demand for future growth. However, this model also has its weakness in that it relies heavily on traffic volume and market stimulation for profitability, because the average yield of tickets is commonly low in the low-cost sector.

To improve this, a number of low-cost airlines have started changing some of the "unique" principles of the business model. Virgin Blue in Australia has started accepting interlining passengers from overseas carriers, bridging the gap between its domestic network and other international networks of overseas partner airlines. This change has brought a certain level of complexity to the low-cost airline operations, so the cost-base of Virgin Blue has risen slightly. However, the gain from connecting the domestic air network of Virgin Blue with overseas partner networks means that there will be a continuous passenger flow between domestic and international networks and this flow also increases the domestic market size as well as the network synergy of Virgin Blue. A similar move has recently been adopted by Jet Blue in the U.S. which partnered with Aer Lingus and Lufthansa to expand domestic and international network coverage on both sides of the U.S. and EU.

Virgin Blue started deploying smaller jets in the Australian domestic market in 2008, besides running the main workhorse of low-cost operations, the B737 family. This move is similar to easyJet's adoption of the A320 along side its fleets of B737, although the use of small Embraer E-jets by Virgin Blue is aimed at expanding small regional markets in Australia. These recent changes reflected the evolution of the classic low-cost carrier business model towards the "new-world carrier" model, as called by Virgin Blue (Thomas 2006). Southwest Airlines has also changed its long standing no-assigned-seats policy, and has started offering some products that may attract more business travellers, hoping that this will improve the yields of tickets and expand its market share in the business travel sector (Field 2007).

6.5.3 Future Philosophy of Airline Scheduling

The context of airline business boils down to two critical facets: *network design* and *a random operating environment*. Network design is both an issue of economics and operations. The pursue of economical gains from a network may inevitably bring negative impacts on the network itself, e.g. complex resource allocation and synchronisation, as well as the creation of pressure on actual operations. The recent trend of robust airline scheduling emerges from the concern of the operational facets of the airline business in a random environment. Ironically, this recent trend has been mostly driven by the financial pressure and on-going cost cutting efforts in the industry since the market downturn in 2001 (9/11 attack), SARS in 2003, and more recently in late 2008 due to the global financial crisis.

Complex airline networks involve hidden costs which are often not realised until the delicate synchronisations among sub-networks are broken. Although simplicity will play a substantial role in shaping future airline networks, some complexities will still exist in most airline networks, regarded as the "necessary evil" in the airline business. Airlines need to strike a good balance between the gains and losses of various network designs, depending on market demands, business strategies and geographical strength. In addition, the balance is also on the operational robustness and reliability of such a network within which stochastic forces play a role in daily operations.

We can sum up the context which the future philosophy of airline business must consider with the following: *airline schedules are pre-planned well ahead of operations, and the operating environment involves random forces which may disrupt schedules and incur extra operating costs in actual operations*. To address this challenge, future airline scheduling philosophy will require strategic planning that considers the possible impacts of stochastic disruptions and the efficiency of operations on schedule plans (Jenner 2009). In the near future, we may see airline scheduling moving along the following two themes: first, *planning for profitability and core businesses*; and second, *planning for robustness and operational flexibility*.

The difference between a network carrier and a low-cost carrier is becoming insignificant; we have seen network carriers adopting some of the effective principles of low-cost business models, e.g. on-demand catering and ticket sales on the internet. We have also seen low-cost carriers adopting some features of network carriers such as interlining services with other carriers and assigning seats to passengers. The "line" between network carriers and low-cost carriers is blurring and it is very likely that hubbing network carriers will co-exist well with point-to-point low-cost carriers in the future.

A key aspect of planning for profitability is to sustain the core businesses of an airline, namely passenger chartering. This core business requires many supporting units within an airline organisation such as ticketing, marketing, planning and sales. Other non-core businesses such as ground handling, catering and maintenance are likely to be outsourced in the future to other airlines, third-party companies, or are to be spun off as independent units/companies within the group of the mother airline organisation. These non-core units do not always work efficiently within an airline structure, especially if the sole function of these units is to serve the core airline business. The recent trend in the industry is to outsource or spin off non-core businesses, so they can become more competitive (i.e. cost efficient) and also produce revenues by serving other airlines and businesses. For instance, Qantas has recently restructured its organisation and moved towards sustaining core units as well as spinning off some non-core units such as catering, JetStar and even its frequent-flyer programme in the near future.

Regarding planning for profitability and core businesses, alliance partners and regional partners play a crucial role in airline scheduling. Although passenger charter is the core business that an airline needs to protect and sustain, some passenger chartering may not be run as cost-efficient as others. For instance, many U.S. carriers are restricted by union bargains on operating regional routes with smaller aircraft. In some other cases, keeping multiple fleet types in an airline may prove to be too costly, because the economies of scale of running multiple fleets cannot always be achieved within an airline. In such a situation, airlines may "outsource" some flight operations to regional partners and meanwhile, save operating costs on running certain fleets (Arnoult 2006). Code-sharing flights with alliance partners can also help in such a situation, providing a cost-efficient way to maintain network coverage and capacity flexibility (Dennis 2005).

The other theme of future airline scheduling is planning for robustness and operational flexibility. Since the capacity of air transport infrastructure is always limited, the greater environment that airlines operate within will function in a similar manner in the future, although capacity constraints at certain busy airports may further constrain the air space. With slow and limited growth in infrastructure capacity, airlines and the industry seek solutions from technology advances to better manage scarce capacity and resources in the system. A key aspect of planning for robustness and flexibility is managing random disruptions and mitigating the impacts on airline operations. In future airline operations, random disruptions will still exist, so airline scheduling has to account for this random force within strategic planning and tactical operation management. Some random events are avoidable (although not entirely) by operational means, e.g. late passenger check-in or missing passengers in terminals, while others are purely random and outside direct control. The consequent impacts due to disruptions can be reduced by delicate strategic schedule planning. The emerging trend of robust airline scheduling is aimed at addressing this issue by bringing stochastic factors from operations back to the drawing board of airline schedule planning.

Robust scheduling and integrated modelling, mentioned earlier in this chapter, are two ways to continue in future airline schedule planning. *Robust scheduling* improves the operational robustness of a network against disruptions by reducing the scale of impacts such as by minimising delay propagation, or by considering possible future disruptions and embedding recovery "options" in a schedule plan. By "internalising" potential disruption costs in the planned costs during schedule planning, a *robust schedule* may incur a higher planned cost after strategic planning, but eventually shall run at a lower *realised cost* level than before.

On the other hand, *integrated modelling* improves the solution quality of various schedule optimisation tasks that have been dealt with individually and sequentially as in the conventional four-stage scheduling paradigm. Through better solution quality, integrated models also reduce the planned costs of airline schedules, and improve the outlook of future profitability due to efficient use of resources and a lower cost base. Although the current focus of integrated modelling is on overcoming mathematical challenges, the future results in Operations Research shall further advance this scheduling forefront. There is a chance that robust scheduling can be combined with the integrated models. Under such a case, further savings can be achieved from *robust integrated modelling* by both enhancing the operational robustness and the solution quality jointly in airline scheduling.

The shift of the airline scheduling paradigm towards robust integrated modelling will take some time to realise in the near future. What was often overlooked in the past in airline scheduling was the realisation of the "context" in which airlines ran the business. Airlines have over-emphasised the gains from complex network operations in the past, and neglected the delicate balance between gains and losses due to network complexity. It seems that more airlines are now willing to trade off commercial gains with operational flexibility and robustness and this trend is encouraging for the airline industry. In addition, the society as a whole will also benefit from this new scheduling philosophy that eventually will provide more efficient and robust air transport services in the future.

References

Abara, J., 1989. Applying integer linear programming to the fleet assignment problem. *Interfaces* 19(4), 20–28.

Abdelghany, K. F., Shah, S. S., Raina, S. and Abdelghany, A. F., 2004. A model for projecting flight delays during irregular operation conditions. *Journal of Air Transport Management* 10, 385–394.

Adeleye, S. and Chung C., 2006. A Simulation based approach for contingency planning for aircraft turnaround operation system activities in airline hubs. *Journal of Air Transportation* 11(1), 140–155.

Adida, E., and Perakis, G., 2006. A robust optimisation approach to dynamic pricing and inventory control with no backorders. *Mathematical Programming* 107(1), 97–129.

Ageeva, J., 2000. Approaches to incorporating robustness into airline scheduling. Master's Thesis, Massachusetts Institute of Technology.

AhmadBeygi, S., Cohn, A., Guan, Y. and Belobaba, P., 2008a. Analysis of the potential for delay propagation in passenger airline networks. *Journal of Air Transport Management* 14(5), 221–236.

AhmadBeygi, S., Cohn, A. and Lapp, M., 2008b. Decreasing airline delay propagation by re-allocating scheduled slack. In Proceedings: The Industry Studies Conference.

Airbus, 2008. Airbus A320 family to benefit from lower maintenance costs. Available from: <http://www.airbus.com/en/presscentre/pressreleases/pressreleases_items/29_nov_04_A320_lower_maintenance_costs.html>, accessed 30 October 2008.

Airports Council International-Europe, 2000. Low-cost airlines bring new blood to airports.

Alderighi, M., Cento, A., Nijkamp, P. and Rietveld, P, 2005. Network competition-the coexistence of hub-and-spoke and point-to-point systems. *Journal of Air Transport Management* 11, 328–334.

Appold, S. J. and Kasarda, J. D., 2006. The appropriate scale of US airport retail activities. *Journal of Air Transport Management* 12, 277–287.

Arguello, M. F., Bard, J. F. and Yu, G., 1998. Models and methods for managing airline irregular operations, in Y. Gang (eds), *Operations Research in the Airline Industry*. Kluwer Academic, Boston.

Arnoult, S., 2006. The hybrid regional. *Air Transport World*, October issue, 64–66.

Ashford, N., Stanton, H. P. M. and Moore, C. A., 1997. *Airport Operations*. McGraw Hill, London.

Australian Government, 2003. Air Navigation Regulations 1947 – made under the Air Navigation Act 1920. Available from: <http://www.comlaw.gov.au/ComLaw/Legislation/LegislativeInstrumentCompilation1.nsf/all/search/A704EF04D084BC5ECA256F710042517F>, accessed 9 January 2009.

Ball, M., Barnhart, C., Nemhauser, G. and Odoni, A., 2007. Air transportation: Irregular operations and control, in C. Barnhart and G. Laporte (eds), *Handbooks in Operations Research and Management Science*, 1–67. Elsevier, Holland.

Barabasi, A. L., 2003. *Linked: How Everything is Connected to Everything Else and What it Means for Business, Science and Everyday Life*. Plume: Massachusetts.

Barnhart, C., Belobaba, P., and Odoni, A. R., 2003. Applications of operations research in the air transport industry. *Transportation Science* 37(4), 368–391.

Barnhart, C., Boland, N. L., Clarke, L. W., Johnson, E.L., Nemhauser, G. L., and Shenoi, R. G., 1998. Flight string models for aircraft fleeting and routing. *Transportation Science* 32(3), 208–220.

Barnhart, C., Farahat, A., and Lohatepanont, M., 2009. Airline fleet assignment with enhanced revenue modelling. *Operations Research* 57(1), 231–244.

Barnhart, C., Kniker, T. and Lohatepanont, M., 2002. Itinerary-based airline fleet assignment. *Transportation Science* 36(2), 199–217.

Bates. J., Polak, J., Jones, P., and Cook, A., 2001. The valuation of reliability for personal travel. *Transportation Research* 37E, 191–229

Bazargan, M., 2004. *Airline Operations and Scheduling.* Ashgate Publishing: Aldershot.

Beatty, R., Hsu, R., Berry, L. and Rome, J., 1998. Preliminary evaluation of flight delay propagation through an airline schedule. Second USA/Europe Air Traffic Management R & D Seminar, Orlando.

Belanger, N., Desaulniers, G., Soumis, F. and Desrosiers, J., 2006. Periodic airline fleet assignment with time windows, spacing constraints and time dependent revenues. *European Journal of Operational Research* 175, 1754–1766.

Ben-Tal, A. and Nemirovski, A., 2000. Robust solutions of linear programming problems contaminated with uncertain data. *Mathematical Programming* 88, 411–421.

Bertsimas, D. and Thiele, A., 2006. A robust optimisation approach to supply chain management. *Operations Research* 54(1), 150–168.

Bhat, V. N., 1995. A multivariate analysis of airline flight delays. *International Journal of Quality and Reliability Management* 12(2), 54–59.

Birge, J. R. and Louveaux, F., 1997. *Introduction to Stochastic Programming.* Springer-Verlag.

Bish, E. K., Suwandechochai, R., and Bish, D. R., 2004. Strategies for managing the flexible capacity in the airline industry. *Naval Research Logistics* 51, 654–685.

Bonini, C. P., Hausman, W. H. and Bierman, H., 1997. *Quantitative Analysis for Management*. 9th edition, Chicago.

Bose, S.K., 2002. *An Introduction to Queueing System*. Academic/Plenum Publishers: New York.

Braaksma, J. P. and Cook, W. J., 1980. Human orientation in transportation terminals. *Journal of Transportation Engineering* 106, 189–203.

Braaksma, J. P. and Shortreed, J. H., 1971. Improving airport gate usage with critical path. *Transportation Engineering Journal of ASCE* 97(2), 187–203.

Bratu, S. and Barnhart, C., 2005. An analysis of passenger delays using flight operations and passenger booking data. *Air Traffic Control Quarterly* 13(1), 1–27.

Bratu, S. and Barnhart, C., 2006. Flight operations recovery: New approaches considering passenger recovery. *Journal of Scheduling* 9(3), 279–298.

British Airport Authority (BAA), 2006. Annual Report. Available from: <http://www.baa.com>, accessed 10 July 2008.

Bureau of Infrastructure, Transport and Regional Economics (BITRE), 2008. Aviation statistics- airline on time performance (February 2008). Available from: <http://www.btre.gov.au/statistics/aviation/otphome.aspx>, accessed 15 May 2008.

Burghouwt, G., 2007. *Airline Network Development in Europe and its Implications for Airport Planning*. Ashgate Publishing: Aldershot.

Button, K., 2002. Debunking some common myths about airport hubs. *Journal of Air Transport Management* 8(3), 177–188.

Chang, K., Howard, K. Oiesen, R., Shisler, L. Tanino, M. and Wambsganss, M. C., 2001. Enhancements to the FAA ground-delay program under collaborative decision making. *Interface* 3, 57–76.

Ching, W.K., 2006. *Markov Chains. Models, Algorithms and Applications*. Springer-Verlag: New York Inc.

Chou, Y. H., 1993a. Airline deregulation and nodal accessibility. *Journal of Transport Geography* 1(1), 36–46.

Chou, Y. H., 1993b. A method for measuring the spatial concentration of airline travel demand. *Transportation Research* 27B(4), 267–273.

Churchill, A., Dada, E., de Barros, A. G. and Wiransinghe, S. C., 2008. Quantifying and validating measures of airport terminal wayfinding. *Journal of Air Transport Management* 14, 151–158.

Civil Aviation Authority (CAA, UK), 1996. Passengers at Birmingham, Gatwick, Heathrow, London City, Luton, Manchester and Stansted Airport in 1996, CAP 677.

Civil Aviation Safety Authority (CASA, Australia), 2008a. Civil Aviation Regulations (CAR) 1988. Available from: <http://www.casa.gov.au>, accessed 4 October 2008.

Civil Aviation Safety Authority (CASA, Australia), 2008b. Civil Aviation Order-Part 20 (CAO-20) Air Service Operations. Available from: <http://www.casa.gov.au/rules/orders/020.htm>, accessed 9 October 2008.

Clark, P., 2001. *Buying the Big Jets*. Ashgate Publishing: Aldershot.

Clarke, J. P., Melconian, T., Bly, E. and Rabbani, F., 2007. MEANS – MIT extensible air network simulation. *Simulation* 83(5), 385–399.

Clarke, L., Johnson, E., Nemhauser, G. and Zhu, Z., 1997. The aircraft rotation problem. *Annals of Operations Research* 69(1), 33–46.

Clarke, M., 2005. Passenger reaccommodation- a higher level of customer service. Presented at AGIFORS Airline Operations Study Group Meeting, Mainz, Germany. Available from: <http://www.agifors.org/>, accessed 19 December 2008.

Clarke, M., Lettovsky, L. and Smith, B., 2000. The development of the airline operations control center, in G. Butler and M. R. Keller (eds), *Handbook of airline operations*. Aviation Week Group of McGraw Hill, Washington D.C.

Cohn, A., and Barnhart, C., 2003. Improving crew scheduling by incorporating key maintenance routing decisions. *Operations Research* 51(3), 387–398.

Cordeau, J. F., Stojkovic, G., Soumis, F. and Desrosiers, J, 2001. Benders decomposition for simultaneous aircraft routing and crew scheduling. *Transportation Science* 35(4), 375–388.

Costa, O. and Paiva, A., 2002. Robust portfolio selection using liner-matrix inequalities. *Journal of Economic Dynamics and Control* 26(6), 889–909.

Dantzig, G. B., 1955. Linear programming under uncertainty. *Management Science* 1(3–4), 197–206.

Dennis, N., 2005. Industry consolidation and future airline network structures in Europe. *Journal of Air Transport Management* 11, 175–183.

Department of Transportation (DoT, U.S.), 2005. Code of Federal Regulations, Title 14- Aeronautics and Space (CFR 14). Available from: <http://www.access.gpo.gov/cgi-bin/cfrassemble.cgi?title=200514>, accessed 9 January 2009.

Department of Transportation (DoT, U.S.), 2008a. Air Consumer Report May 2008. Available from: <http://airconsumer.ost.dot.gov/reports/atcr08.htm>, accessed 15 May 2008.

Department of Transportation (DoT, U.S.), 2008b. Revision of airline service quality performance reports and disclosure requirements. Available from: <http://www.bts.gov/laws_and_regulations/docs/part234_4cy2008.html>, accessed 15 May 2008.

Desaulniers, G., Desrosiers, J., Dumas, Y., Soloman, M. M., and Soumis, F., 1997. Daily aircraft routing and scheduling. *Management Science* 43(6), 841–855.

Dobbyn, T., 2000. US prepares plan to cut summer flight delays. Reuters Wire, 3 March 2000.

Doganis, R., 2002. *Flying Off Course*. Routledge Publishing: New York.

Doganis, R., 2005. Harsh realities. *Airline Business*, November issue, 77–81.

DRI – WEFA Inc. in collaboration with The Campbell-Hill Aviation Group Inc., 2002. The National Economic Impact of Civil Aviation.

Ehrgott, M. and Ryan, D. M., 2002. Constructing robust crew schedules with Bicriteria optimisation. *Journal of Multi-Criteria Decision Analysis* 11, 139–150.

Eilstrup, F., 2000. Determining the absolute minimum turnaround time for your flights – what is realistically achievable? European Carrier Perspective. Proceedings of Minimising Aircraft Turnaround Times Conference, London England.

Erdmann, A., Nolte, A. Noltemeier, A. and Schrader, R., 2001. Modeling and solving an airline schedule generation problem. *Annals of Operations Research* 107, 117–142.

Eurocontrol, 2004a. Evaluating the true cost to airlines of one minute of airborne or ground delay. Performance Review Commission.

Eurocontrol, 2004b. Delays to air transport in Europe: Annual report.

Eurocontrol, 2005. Delays to air transport in Europe: Annual report.

Eurocontrol, 2007. Delays to air transport in Europe: Annual report.

Eurocontrol Experimental Centre, 2002. Analysis of passengers delays: an exploratory case study. EEC Note No. 10/02 Project PFE-F-FM.

European Union, 2005. Air passenger rights. Available from: <http://ec.europa.eu/transport/air_portal/passenger_rights/information_en.htm>, accessed 12 December 2008.

Field, D., 2005. Hubs attract low-cost trade. *Airline Business*, March issue, 15.

Field, D., 2007. Southwest's model shift. *Airline Business*, December issue.

Fishman, G. S., 1997. *Monte Carlo: Concepts, Algorithms, and Applications*. Springer: New York.

Francis, G., Fidato, A. and Humphreys, I., 2003. Airport-airline interaction: the impact of low-cost carriers on two European airports. *Journal of Air Transport Management* 9, 267–273.

Gao, C., Johnson, E. L., and Smith, B., 2009. Integrated airline fleet and crew robust planning. *Transportation Science* 43(1), 2–16.

Ghobrial, B. and Kanafani, A., 1995. Future of airline hubbed networks: some policy implications. *Journal of Transportation Engineering* 121(2), 124–134.

Gilbo, E. P., 1993. Airport capacity: representation, estimation, optimisation. *IEEE Transactions on Control Systems Technology* 1(3), 144–153.

Gillen, D. and Lall, A., 2004. Competitive advantage of low-cost carriers: some implications for airports. *Journal of Air Transport Management* 10(1), 41–50.

Gittell, J. H., 1995. Cost/quality trade-offs in the departure process? Evidence from the major U.S. airlines. *Transportation Research Record* 1480, 25–36.

Gittell, J. H., 2001. Supervisory span, relational coordination and flight departure performance: a reassessment of post-bureaucracy theory. *Organization Science* 12(4), 468–490.

Goedeking, P. and Sala, S., 2003. Breaking the bank. *Airline Business*, September issue, 93-97.

Goldnadel, F., 2007. Presentation of the CDG2 Hub. Available from: <http://www.aeroportsdeparis.fr/Adp/Resources>, accessed 3 April 2009.

Goolsbee, A., 2005. Tragedy of the airport. Available from: <http://slate.msn.com/id/2123240>, accessed 11 February 2009.

Green, D. G. and Bransden, T. G., 2006. Complexity theory, in: *McGraw-Hill Encyclopaedia of Science and Technology*, 507–511. McGraw-Hill: New York.

Hane, C. A., Barnhart, C., Johnson, E. L., Marsten, R. E., Nemhauser, G. L. and Sigismondi, G., 1995. The fleet assignment problem: solving a large-scale integer program. *Mathematical Programming* 70(2), 211–232.

Haouari, M., Aissaoui, N. and Mansour, F. Z., 2009. Network flow-based approaches for integrated aircraft fleeting and routing. *European Journal of Operations Research* 193(2), 591–599.

Hassounah, M. I., and Steuart, G. N., 1993. Demand for aircraft gates. *Transportation Research Record* 1423, 26–33.

Hensher, D. A., Rose, J. M. and Greene, W. H., 2005. *Applied Choice Analysis: a Primer.* Cambridge University Press: Cambridge.

Holloway, S., 2008. *Straight and Level: Practical Airline Economics*. Ashgate Publishing: Aldershot.

Hsu, C. and Shih, H., 2008. Small-world network theory in the study of network connectivity and efficiency of complementary international alliances. *Journal of Air Transport Management* 14, 123–129.

Hsu, C.I. and Wen, Y.H., 2000. Application of grey theory and multiobjective programming towards airline network design. *European Journal of Operational Research* 127(1), 44–68.

Hsu, C.I. and Wen, Y.H., 2003. Determining flight frequencies on an airline network with demand-supply interactions. *Transportation Research* 39E, 417–441.

Huang, S.C., 1998. Airline delay perturbation problem under minor disturbance. NEXTOR working paper (WP-98-1), University of California at Berkeley.

International Air Transport Association (IATA), 1997. *Airport Handling Manual.* 17th Edition.

International Air Transport Association (IATA), 2003. *Airport Handling Manual.* 23rd Edition.

International Air Transportation Association (IATA), 2004. *Worldwide Scheduling Guideline.*

IATA, 2006. *Worldwide Scheduling Guidelines*. 13th Edition.

IATA, 2008. Radio frequency ID (RFID) for aviation. Available from: <http://www.iata.org/stbsupportportal/rfid/>, accessed 20 May 2008

Institut du Transport Aérien, 2000. Costs of air transport delay in Europe.

International Civil Aviation Organisation (ICAO), 1997a. *Digest of Statistics, Series F. Financial Data Commercial Air Carriers.*

International Civil Aviation Organisation (ICAO), 1997b. *Digest of Statistics, Series F, Fleet-Personnel.*

Ivy, R. J., 1993. Variations in hub service in the US domestic air transportation network. *Journal of Transport Geography* 1(4), 211–218.

Jacobs, T. L., Smith, B. C. and Johnson, E. L., 2008. Incorporating network flow effects into the airline fleet assignment process. *Transportation Science* 45(4), 514–529.

Janic, M., 1997. The flow management problem in air traffic control: A model of assigning priorities for landings at a congested airport. *Transportation Planning and Technology* 20, 131–162.

Jarrah, A., Goodstein, J. and Narasimhan, R., 2000. An efficient airline re-fleeting model for the incremental modification of planned fleet assignments. *Transportation Science* 34(4), 349–363.

Jarrah, A. I. Z. and Yu, G., 1993. A decision support framework for airline flight cancellations and delays. *Transportation Science* 27(3), 266–280.

Jenner, G., 2009. Planning ahead. *Airline Business*, April issue, 38–40.

Kang, L., 2004. Degradable airline scheduling: An approach to improve operational robustness and differentiate service quality. PhD Dissertation, Massachusetts Institute of Technology.

Kim, W., Park, Y. and Kim, B. J., 2004. Estimating hourly variations in passenger volume at airports using dwelling time distributions. *Journal of Air Transport Management* 10, 395–400.

Klabjan, D., 2005. Large-scale models in the airline industry, in M. M. Solomon, G. Desaulniers, and J. Desrosiers (eds), *Column Generation*, 163–195. Springer: New York.

Klabjan, D., Johnson, E. L., and Nemhauser, G. L., 2001. Solving large airline crew scheduling problems: Random pairing generation and strong branching. *Computational Optimization and Applications* 20, 73–91.

Klabjan, D., Johnson, E. L., Nemhauser, G. L., Gelman, E. and Ramaswamy, S., 2002. Airline crew scheduling with time windows and plane count constraints. *Transportation Science* 36, 337–348.

Klafehn, K., Weinroth, J., and Boronico, J., 1996. *Computer Simulation in Operations Management*. Quorum: Westport.

Lan, S., Clarke, J., and Barnhart, C., 2006. Planning for robust airline operations: optimising aircraft routings and flight departure times to minimise passenger disruptions. *Transportation Science* 40(1), 15–28.

Lawton, T. C., 2002. *Cleared for Take-Off.* Ashgate Publishing: Aldershot.

Lederer, P.J. and Namimadom, R. S., 1998. Airline network design. *Operations Research* 46(6), 785–804.

Lee, K., and Moore, J., 2003. On time performance. AGIFORS Airline Operations Conference, Auckland, New Zealand. Available from: <http://www.agifors.org>, accessed 11 February 2009.

Lee, L. H., Lee, C. U. and Tan, Y. P., 2007. A multi-objective genetic algorithm for robust flight scheduling using simulation. *European Journal of Operational Research* 177, 1948–1968.

Lettovsky, L, Johnson, E. L. and Nemhauser, G. L., 2000. Airline crew recovery. *Transportation Science* 5, 232–255.

Levin, A., 1971. Scheduling and fleet routing models for transportation systems. *Transportation Science* 5, 232–255.

Lohatepanont, M. and Barnhart, C., 2004. Airline schedule planning: integrated models and algorithms for schedule design and fleet assignment. *Transportation Science* 38, 19–32.

Luo, S. and Yu, G., 1997. On the airline schedule perturbation problem caused by the ground delay program. *Transportation Science* 31(4), 298–311.

Makri, A., and Klabjan, D., 2004. A new pricing scheme for airline crew scheduling. *INFORMS Journal on Computing* 16(1), 56–67.

Malighetti, P., Paleari, S. and Redondi, R., 2008. Connectivity of the European airport network: 'Self-help hubbing' and business implications. *Journal of Air Transport Management* 14(2), 53–65.

Markowitz, H. M., 1952. Portfolio selection. *Journal of Finance* 7(1), 77–91.

Markowitz, H. M., 1959. *Portfolio Selection.* Wiley: New York.

Martin, J. C. and Voltes-Dorta, A., 2008. Theoretical evidence of existing pitfalls in measuring hubbing practices in airline networks. *Network and Spatial Economics* 8, 161–181.

Mederer, M., and Frank, M., 2002. Increasing robustness of flight schedules through stochastic modeling of planning parameters. AGIFORS Airline Operations Conference. Available from: <http://www.agifors.org>, accessed 11 February 2009.

Mercier, A., Cordeau, J. F. and Soumis, F., 2005. A computation study of Benders decomposition for the integrated aircraft routing and crew scheduling problem. *Computational Operations Research* 32(6), 1451–1476.

Mercier, A. and Soumis, F., 2007. An integrated aircraft routing, crew scheduling and flight retiming model. *Computers and Operations Research* 34, 2251–2265.

NASA, 2008. Virtual airspace modeling and simulation (VAMS) project. Available from: http://vams.arc.nasa.gov/, accessed 10 June 2008.

Ndoh, N. N. and Ashford, N., 1993. Evaluation of airport access level of service, *Transportation Research Record* 1423, 34–39.

Nyquist, D. C. and McFadden, K. L., 2008. A study of the airline boarding problem. *Journal of Air Transport Management* 14, 197–204.

Oba, K., 2007. Supply side revenue management: ANA's experience with an optimised equipment swap system. Scheduling and Strategic Planning Study Group, AGIFORS. Available from: <http://www.agifors.org>, accessed 20 March 2009.

Papadakos, N., 2009. Integrated airline scheduling. *Computers and Operations Research* 36(1), 176–195.

Papatheodorou, A. and Lei, Z., 2006. Leisure travel in Europe and airline business models: A study of regional airports in Great Britain. *Journal of Air Transport Management* 12, 47–52.

Park, Y. H. and Ahn, S. B, 2003. Optimal assignment for check-in counters based on passenger arrival behavior at an airport. *Transportation Planning and Technology* 26, 397–416.

Pauley, G.S., Ormerod, R.J., Woolsey, R. and Talluri, K., 1998. The four-day aircraft maintenance routing problem. *Transportation Science* 32(1), 43–53.

Rexing, B., Barnhart, C., Kniker, T., Jarrah, A. and Krishnamurthy, N., 2000. Airline fleet assignment with time windows. *Transportation Science* 34(1), 1–20.

Reynolds-Feighan, A., 1998. The impact of U.S. airline deregulation on airport traffic patterns. *Geographical Analysis* 30(3), 234–253.

Richetta, O. and Odoni, A. R., 1993. Solving optimally the static ground-holding policy problem in air traffic control. *Transportation Science* 27(3), 228–238.

Rosenberger, J., Johnson, E. L. and Nemhauser, G. L., 2003. Rerouting aircraft for airline recovery. *Transportation Science* 37, 408–421.

Rosenberger, J., Johnson, E. L. and Nemhauser, G. L., 2004. A robust fleet assignment model with hub isolation and short cycles. *Transportation Science* 38(3), 357–368.

Rosenberger, J., Shaefer, A. J., Goldsman, D., Johnson, E. L., Kleywegt, A. J. and Nemhauser, G. L., 2002. A stochastic model of airline operations. *Transportation Science* 36(4), 357–377.

Ryan, D., 1992. The solution of massive generalized set partitioning problems in aircrew rostering. *Journal of Operational Research Society* 43(5), 459–467.

Sabre, 2008. Sabre airline solutions- AirOps suite. Available from: <http://www.sabreairlinesolutions.com/products/operate/airops.htm>, accessed 5 October 2008.

Sandhu, R., 2007. Station specific aircraft turn times and applications for schedule reliability. AGIFORS, Tokyo. Available from: <http://www.agifors.org>, accessed 11 February 2009.

Sandhu, R. and Klabjan, D., 2007. Integrated airline fleeting and crew-pairing decisions. *Operations Research* 55(3), 439–456.

Sarmadi, S., 2004. Minimizing airline passenger delay through integrated flight scheduling and aircraft routing. Master's Thesis, Massachusetts Institute of Technology. Available from: <http://hdl.handle.net/1721.1/29401>, accessed 24 March 2009.

Schaefer, A. J., Johnson, E. L., Kleywegt, A. J. and Nemhauser, G. L., 2005. Airline crew scheduling under uncertainty. *Transportation Science* 39, 340–348.

Schiewe, J., 2005. OCC forecaster. AGIFORS Airline Operations Conference. Available from: <http://www.agifors.org>, accessed 11 March 2009.

Schiphol Group, 2007. Annual Report. Available from: <http://schiphol.com>, accessed 10 July 2008.

Shaw, S., 1993. Hub structures of major US passenger airlines. *Journal of Transport Geography* 1(1), 47–58.

Shebalov, S. and Klabjan, D., 2006. Robust airline crew scheduling: move-up crews. *Transportation Science* 40, 300–312.

Sherali, H. D., Bish, E. K. and Zhu, X., 2006. Invited Tutorial: Airline fleet assignment concepts, models, and algorithms. *European Journal of Operational Research* 172, 1–30.

Sherali, H. D. and Zhu, X., 2008. Two-stage fleet assignment model considering stochastic passenger demands. *Operations Research* 56(2), 383–399.

Shifrin, C., 2004. That's entertainment. *Airline Business*, January issue, 46–47.

Smith, B. C., 2004. Robust airline fleet assignment. PhD Dissertation. Georgia Institute of Technology, Atlanta GA, USA.

Smith, B. and Johnson, E. L., 2006. Robust airline fleet assignment: Imposing station purity using station decomposition. *Transportation Science* 40, 497–516.

Steffen, J. H., 2008. Optimal boarding method for airline passengers. *Journal of Air Transport Management* 14, 146–150.

Suzuki,Y., 2000. The relationship between on-time performance and airline market share: a new approach. *Transportation Research* 36E, 139–154.

Sydney Airport, 2007. Annual Report. Available from: <http://www.sydneyairport.com.au/SACL/default.htm>, accessed 10 July 2008.

Taha, H. A., 1992. *Operations Research: An Introdution.* Macmillian Publishing: New Jersey.

Teodorovic, D., 1983. Flight frequency determination. *Journal of Transportation Engineering* 109(5), 747–757.

Teodorovic, D., Kalic, M., and Pavkovic, G., 1994. The potential for using fuzzy set theory in airline network design. *Transportation Research* 28B(2), 103–121.

Teodorovic, D. and Krcmar-Nozic, E., 1989. Multicriteria model to determine flight frequencies on an airline network under competitive conditions. *Transportation Science* 23(1), 14–25.

Teodorovic, D. and Stojkovic, G., 1990. Model for operational daily airline scheduling. *Transportation Planning and Technology* 14, 273–285.

Teodorovic, D. and Stojkovic, G., 1995. Model to reduce airline schedule disturbances. *Journal of Transportation Engineering* 121, 324–331.

Thomas, G., 2006. Taking care of business. *Air Transport World*, October issue, 54-58.

Thon, A., 2005. ALLEGRO SERVICES make online control visible. AGIFORS Airline Operations Conference. Available from: <http://www.agifors.org>, accessed 15 March 2009.

Torres, E., Dominguez, J. S., Valdes, L. and Aza, R., 2005. Passenger waiting time in an airport and expenditure carried out in the commercial area. *Journal of Air Transport Management* 11, 363–367.

Tosic, V., Babic, O., Cangalovic, M. and Hohlacov, D., 1995. Some models and algorithms for en-route air traffic flow management. *Transportation Planning and Technology* 19, 147–164.

Tosic, V. and Basic, O., 1984. Quantitative evaluation of passenger terminal orientation. *Journal of Advanced Transportation* 18, 279–295.

Trietsch, D., 1993. Scheduling flights at hub airports. *Transportation Research* 27(B), 133–150.

Tutuncu, R. H. and Koenig, M., 2004. Robust asset allocation. *Annals of Operations Research* 132, 157–187.

Valtat, A., 2007. Eurocontrol legal framework. Available from: <http://www.eurocontrol.int/corporate/public/standard_page/org_legal_framework.html>, accessed 13 January 2009.

Van den Briel, M. H. L., Villalobos, J. R., Hogg, G. L., Lindenmann, T. and Mule, V. A., 2005. America West airlines develops efficient boarding strategies. *Interfaces* 35, 191–201.

Vance, P., Barnhart, C., Johnson, E. L. and Nemhauser, G. L., 1997. Airline crew scheduling: A new formulation and decomposition algorithm. *Operations Research* 45(2), 188–200.

Weide, O., Ryan, D. and Ehrgott, M., 2007. Iterative airline scheduling. University of Auckland School of Engineering. Available from: <http://hdl.handle.net/2292/2669>, accessed 24 March 2009.

Wikipedia, 2008a. Aircraft communication addressing and reporting system (ACARS). Available from: <http://en.wikipedia.org/wiki/Aircraft_Communication_Addressing_and_Reporting_System>, accessed 15 May 2008.

Wikipedia, 2008b. Program evaluation and review technique. Available from: <http://en.wikipedia.org/wiki/Program_Evaluation_and_Review_Technique>, accessed 23 June 2008.

Watterson, A. and De Proost, D., 1999. On-time performance: operational challenges for airlines. Available from: <http://www.mercer.com>, accessed 20 June 2005.

Watts, D. J. and Strogatz, S. H., 1998. Collective dynamics of 'small-world' networks. *Nature* 393, 440–442.

Wu, C. L., 2005. Inherent delays and operational reliability of airline schedules. *Journal of Air Transport Management* 11, 273–282.

Wu, C. L., 2006. Improving airline network robustness and operational reliability by sequential optimisation algorithms. *Journal of Network and Spatial Economics* 6 (3–4), 235–251.

Wu, C. L., 2008. Monitoring aircraft turnaround operations – framework development, application and implications for airline operations. *Transportation Planing and Technology* 31(2), 215–228.

Wu, C. L. and Caves, R. E., 2000. Aircraft operational costs and turnaround efficiency at airports. *Journal of Air Transport Management* 6, 201–208.

Wu, C. L. and Caves, R. E., 2002. Modelling of aircraft rotation in a multiple airport environment. *Transportation Research* 38E, 265–277.

Wu, C. L. and Caves, R. E., 2004. Modelling and simulation of aircraft turnaround operations at airports. *Transportation Planning and Technology* 27(1), 25–46.

Yan, S. Y. and Yang, D. H., 1996. A decision support framework for handling schedule perturbation. *Transportation Research* 30B(6), 405–419.

Yan, S. and Young, H. F., 1996. A decision support framework for multi-fleet routing and multi-stop flight scheduling. *Transportation Research* 30A, 379–398.

Yen, J. W. and Birge, J. R., 2006. A stochastic programming approach to the airline crew scheduling problem. *Transportation Science* 40(1), 3–14.

Yu, G., Arguello, M., Song, G., McCowan, S. and White, A., 2003. A new era for crew recovery at Continental Airlines. *Interfaces* 33, 5–22.

Zhao, W., Morley, S. and Srinivasan, B., 2007. Flexible capacity at United Airlines. Scheduling and Strategic Planning Study Group, AGIFORS. Available from: <http://www.agifors.org>, accessed 20 March 2009.

Index